FLOWERS AND VEGETABLES OF INDIA

Romesh Kumar Sud
and
Sudhir Kumar

SCIENTIFIC
PUBLISHERS

Published by:
Scientific Publishers (India)
5 A, New Pali Road, P.O. Box 91
Jodhpur - 342 001 (India)

Branch Office
Scientific Publishers (India)
4806/24, Ansari Road, Daryaganj
New Delhi - 110 002 (India)

E-mail: info@scientificpub.com
Website: www.scientificpub.com

Print: 2017

ISBN: 978-81-7233-361-4

Lasertype Set: Rajesh Ojha
Printed in India

PREFACE

India is a country of very many climates at one and the same season. Summers are very pleasant and cool in the Northern Hill areas of Jammu and Kashmir, Himachal Pradesh, Uttaranchal very much akin to the temperate conditions of European countries. Winter season in the Gangetic plains and also in Peninsular India is very pleasant, thereby providing conditions for culture of varied types of fruits and vegetables. After Independence, India has made very many big strides in every sphere of Agriculture distinctions, firstly by becoming self-sufficient in grains and then getting into the white revolution. India with its vegetable production stands next to China in the world. In the sphere of flower culture, India has not gained any popularity in the world markets because of its direction only for domestic needs of flowers and un-awareness about the so-called cut flowers. Also Indian technology is not that ahead and advanced in covered crop culture as that of the western world. Also Indian farmers are very much un-aware of the continental vegetables which are much in demand both in the domestic and export markets. These continental vegetables are very high value crops and earn huge returns per unit area.

Thrust of the developmental activities in agriculture has been now directed to the cut flower, potted plant and nursery raising for ornamentals and culture of high value crops. This thinking and direction will remain incomplete till there is full acquaintance of the farmer with the inherent techniques of production, post harvest technology, marketing both in the domestic and export markets and getting him closer to the knowledge of understanding the export methodology required for generating good incomes from his produce of flowers and vegetables by adopting the latest technologies. All the Agricultural Universities and Research Institutes involved in this aspect have developed new flower and vegetable varieties suited for Indian conditions and much in demand in the International markets, are not enough steps, till all the concerned in this technology, may be students, Research Scientists and the farmers or the Bureaucrats are made aware of the inherent practical needs of the day.

The presented manuscript in the book form on "Flowers and Vegetables of India", has taken all the due care in compiling the details on the cut flower culture both in the open and also under covers and acquainting the user and the reader with the export entry into the market and its requirements, transportation under controlled conditions which are the top most requirements have been given. Also details of establishing own Nurseries, Tissue Culture Laboratories and Project Costs on Cut flower Floriculture have been given in a very easy to understand way, so that the entrepreneur can prepare his own estimates and project planning details for gaining access to the Government requirements for the release of grants / subsidies and loan funds in a very easy fashion. Organic culture which is likely to become a part of the farmer and the entrepreneur in agriculture has also been given in a manner that, nothing is left for the future reference of the subject. In the Vegetable crop culture, detailed information on the history, botany and the vegetable family as a whole has been tried to be presented in such a way that the reader acquires all the best possible knowledge from Origin to different species and the relevance of cultivation with regard to demands of the country and the continent. In this direction differences between every known crop has been given, as in the case of Peas, we, in India are aware of only Shelled Peas and not the Snow and Snap Peas. The same way, Beans are known only as green vegetable and not as Shelled beans. Thus, the reader has been kept to be at the best receiving end and to acquaint him all about the inherent techniques of creating awareness and educating in a simple manner for practice in the field. Success stories of the people from Taiwan and other countries are the motivating stones of this book. This Book thus is an intentional write-up for the students and rural educated folks interested in any vocation of agriculture, whether it is Floriculture or Olericulture.

Apart from the above, reasons for un-awareness of the farmer are there which will become the guide stones to achieve the desired results in the line once he thinks to adopt it. The Book in its present form is a consequence of the practical experience gained by the authors in the field for culture, post harvest and marketing of fresh products and processing for food security. Contents contained for different cultural, industrial and marketing aspects of crops can usher into a new life for the village. It is very high time that farmer understands the intrigues of modern crop culture through his own vision and translates with his untiring efforts. Greatest emphasis in the present book has been laid on the competitiveness of India with other world countries, where India stands to gain at every front whether cheaper production costs or technology understanding or geographical location of India in the center of the world export trade, are very vital.

New Delhi **AUTHORS**

CONTENTS

Part – II

CONTINENTAL VEGETABLES

PART - I
FLORICULTURE

Botanical Name	Local/Common/English Name
Brassica nigra	Black / Brown Mustard
Brassica juncea	Brown Mustard, Rai, Indian Mustard
Brassica rugosus	Bela Rape
Brassica alba syn. *B. hirta* syn. *Sinapsis alba*	White or yellow Mustard
Brassica besseriana	Sarepta Mustard / Asl-Rai, Brown Mustard
Brassica carinata	Ethiopian Rape seed/Abyssinian Mustard
Brassica campestris	
Brassica rapa	
Brassica japonica	
Brassica napus	

Plant : is a perennial herb but grown as annual or bi-ennial, erect with long branch's with tap root system. Flowers appear in racemes. Plant height varies for each specie and ranges between 2-6 ft. Plant is fairly resistant to droughts.

Climate : Hardy cool season plant, grows best in conditions of monthly average temperature of [illegible]. Thrives best in areas with hot days and cool nights. Rainfall of 3.5-17 cms is quite sufficient for good crop. Best growth temperature for all sarson species is 5.5 - 27°C.

Soil : Culture of Brassicas suits very much for light, sandy loam, deep rich fertile soils. Grows in soils with a very wide range of pH of 4.5 - 8.2.

Seed Rate : is generally the same in all the species which ranges between 1.6-2.0 Kgs/acre. In Britain the seed rate for heavy and light soils are 2.8 and 1.8 Kgs/acre. Seeds are sown maintaining a plant to plant distance between 4-6 inches with a row to row distance of 6-20 [illegible].

Sowing Time : starts with the onset of winters in India and spring in Europe. Certain varieties have been bred with very short period of 90 days maturity, which have a special place in the Crop rotation for having 2 crops in the same season thereby also increasing the crop intensity. Culture of Toria before the main crop of Wheat is one such example. Sarson is also sown as a mixed crop in the rainfed areas to have the benefit of weather vagaries and avoid crop failure.

CHAPTER - 1

WORLD FLORICULTURE

WORLD LEADERS IN FLORICULTURE

In Floriculture industry Holland, USA, Israel with a share of 60%, 10%, 4% in Floriculture trade are considered leaders. Technology development of these leaders is inherent in their integrated approach to Floriculture i.e cultivation of Flowers both in the open and also under cover conditions, breeding of Flowers for high yields, colourful, eye catching varieties, post harvest management to increase shelf life of the flowers are some of the best achievements. Apart from Israel other non-European leaders are Colombia, Kenya, Ecuador, Zimbabwe and Thailand in Floriculture.

NETHERLAND: is a dominant player of International Floriculture trade with its ranking as 5th largest importer after Germany, USA, UK and France. Netherlands is also by far the largest exporter of cut flowers in the world. Exports worth 2.15 billion USD accounted for 64% of world exports in cut flowers. Per capita consumption of flowers is also high and was 83 Guilders. With its ranking as the largest producer of cut flowers has got distinction of concentrated areas in the industry. These concentrated areas are Aalsmeer in Amsterdam and Hague. Bulk of Floriculture nearly 60% is done under covers and it was 3669 Ha in 1995, which was 18.96% more than the open culture spread in an area of 2499 Ha.

In order of Floriculture specie rankings Roses, Chrysanthemums, Carnations, Lilies, Orchids, Gerberas, Alstroemeria, Gypsophylla, Anthuriums, Nerines followed one after the other. Roses occupied 25% of the total area.

There are 9 Flower auction centers in Holland and the more famous ones in these are "ALSMEER FLOWER AUCTION (VBN)" and "NAALDWIJK (BVH)" and "RIJNSBURG" which accounted for 95% auction trade. TELE FLOWER CENTER (TFA) is the only private owned center for auction of flowers in Holland. 95% sales pass through the auction clock system. The qualitative difference in these auctions is that all players of the trade whether be auctioneers or Importers or Major buyers all require pre-treated flowers for longer shelf life.

COLOMBIA : After Netherlands in the European Union, Colombia in South American region is the new leader in Floriculture second to Netherlands in the export trade. Colombian farmer and the corporate by virtue of their stead fastness in world market demand study and dedicated efforts for this sector has achieved this position. Presently more than 30,000 Flower box's are shipped per day to different destinations in the world. Main market for the Colombian flowers are North American countries USA, Canada, Holland, Germany, Spain and United Kingdom in the European Union. New markets accessed by Colombian marketers are Japan, East Europe, Finland, Brazil, Argentina, Chile and Venezuela.

With a total of 4500 ha flower plantations export earnings from this industry was 556.2 million US$ during 1998 and a further generation of employment to the tune of 1,30,000 direct and indirect jobs for the Colombians, was a creditable achievement for one sector. Floraculture International reports that Colombian floriculture has so good an impact that out of every 2 flowers sold in the market second flower is Colombian. Impact of Floriculture industry development is so great in the economy and culture of Colombia that it is a comfortable occupation for all the Columbians and country as a whole.

BRAZIL : Brazilian floriculture is in constant state of modernization as evidenced by the presence of Ultra modern Green Houses well equipped with heating, pad and fan cooling, shading screens and rolling bench's being the requirement of protected or covered crop culture. Other aspects of bright future of Brazilian floriculture are low labour costs, stable inflation rate, competitive Air freight structure, lucrative domestic markets and favourable climate. Brazil has 4500 Ha area in floriculture with 3600 growers in this industry. Production value of the floriculture products in the wholesale market is 180 million US$ which in the retail sale is 1200 million US$.

SPAIN : Area under floriculture crops in Spain is 2000 Ha. Value earning is 572 million US$. Main cultural areas in the country are Andaluci region, Galicia, Murcia which account for 60%, 15%, and 12% respectively. Dianthus is the main flower crop for export. Dianthus is produced in 1250 Ha area and exports to European countries - Netherlands stands at 70%, UK-17%, France 12% during the year 2001 (Floraculture International). For the benefit and education of the local farmers in floriculture CIFA – Center for Agriculture Research and Education Training located at Chipiona near Cadiz in Andalucia caters to technical support for the farmers.

TURKEY : Turkey is a small country of the Middle East and has a good share in flower exports even with very small area. During the year 1992-93 area was 669 Ha, which got reduced to 557 Ha in the year 1997-98. Area under outdoor ornamental plant culture was only 53 Ha which rose to 1010 ha by the year 1998. Cut flower exports from Turkey was 13.53 million US$ during the year 1998.

OTHER WORLD COUNTRIES IN FLORICULTURE

African Continent Countries : As Latin Americas floriculture is geared to N. America, African exporters are similarly aiming European markets. Kenya and Zimbabwe are the two floriculture developed countries of the continent without ACP status. Other countries of Africa which qualify for ACP Trade Treaty are Tanzania, Uganda, Zambia. South Africa with its total production has 75% domestic demand and rest for export. Present status of growing floriculture in Africa is due to grower initiative and their capacity to invest and take risks. Along with this growers have further built in their capability for quality professionalism in floriculture export trade. These six countries of the African continent rank among the 11 European Union suppliers. Kenya has overtaken Israel and Columbia in 1999 floriculture export. Countries like Tanzania, Uganda, S. Africa are there in the race of Floricultural products, but are not reckoned as world leaders.

KENYA : From the African continent Kenya has proved its worth with its long proven ability as one of the leading country in that continent for flower export. During the year 2001, it exported

41,400 tins of flowers, which was an increase of 36.9% over the years 1998 and 2000. Kenya production of floral products has a share of 65% for Rose cut flower. Achievement today in Kenya is categorized to the active participation of the private entrepreneur and his investment attitude.

ZIMBABWE : In Africa after Kenya it is the Zimbabwe, standing at No. 2 for export of flowers to the EU countries. In the year 2000, it exported nearly 2000 tons. Floral product production in Zimbabwe's total area of 1150 Ha with 300 growers comprises of Roses, Protea and summer flowers. Main markets covered are Netherlands 87%, Germany 4.5% and South Africa 5.5%.

ZAMBIA : With a steady growth in its exports Zambia's exports of floral products during the year 2001 were to the tune of 3764 tons. In this production Roses accounted for 95% production. Zambia with its 31 growers and an area of 195 Ha. Fetched a return of 29 million US$. The trade growth was + 10%. Markets covered were Netherlands 95 % and Germany 5%.

AUSTRALIA : In the Oceania region Australian cut flower industry has only a share of 1% in the total world trade. Main export range of flowers and the potted plants include Wax Flower, Kangaroo Paw, Protea, Banksia, Thryptomene, Stirlingia, Leucadendron. The export of the Australian products is from the selection of the wild plants of the continent which conformed to the customer likes in the world trade.

TAIWAN : In the Far East Asia Taiwan is one name which is getting a reckoning in the world trade of Floriculture. In the initial years of the formation and coming up of Taiwan, endeavours were focused on the production of processed Pineapples, Silk, Asparagus and canned Mushrooms. These activities generated an enormous amount of foreign investment in Taiwan and established an era of Industrial development nourished by the support of agricultural production. With this Taiwan faced to the International demands and competitions and challenges to the Floriculture and agriculture in a unique way of competing with the situation. Result was that Taiwan today is placed as the largest exporter country of Phaleonopsis, earning it a name and fame in the world as "Kingdom of Orchids".

Basic strength of Taiwan lies in its determination to achieve the objectives. Though many of Taiwan's Orchid growers have no formal horticultural training, but they are strong and self motivated contenders and learners. This starts from the interest for orchid production as a learner. With the passage of time Taiwanese have developed their floriculture from individual growers to partnerships and then corporation invested companies typical of Taiwan. Success stories of farmers like Chiang-Kuei Feng of SoGo Orchid Nursery and Mr. Fon are a few which has shown the way for other Orchid growers.

Table : Showing African Countries Floriculture Status

	COUNTRIES					
	Kenya	Zimbabwe	Uganda	Tanzania	Zambia	S. Africa
No. of Growers	140	300	20	9	31	900
Area in Ha.	2180	1150	127	85	195	1900
Export in Tons	41396	20000	3710	2478	3764	-
Type of item:						
• Roses	30253	-	3069	1813	-	-
• Protea	-	-	-	-	-	-
• Young Cuttings	1339	-	-	612	-	-
• Others	9804	-	-	9804	-	-
• Netherlands	90	87	90	80	95	56
• U. Kingdom	4	-	4	-	-	-
• Belgium	3	-	3	-	-	-
• Germany	-	4.5	-	19	3	23
• N. America	-	-	-	-	-	11.4
• Asia/ Australia	-	-	-	-	-	12.8

(**Source** : Floraculture – June, 2002)

Another success story is that of Louis Stacy a former Navy Fighter Pilot turned an experienced Floriculture entrepreneur of the USA for the past 30 years. He specialized only with the Pansies. Today Stacy is one of the leading colour growers in USA with 29 Ha of growing area, 4 ha of shade houses and 3.8 ha of Green houses. The nursery production operations spanned in three locations. Stacy's advice to the new comers in the field of floriculture is – to do things in a way as they desire to be done and to work as a team. Prevent problems rather than waiting for the problems to happen. Do not worry, what other are doing, only worry what you are doing.

CHAPTER - 2

FLORICULTURE IN INDIA

INTRODUCTION

Flowers are a symbol of softness and tenderness and every one loves the beauty of flowers, their form, colour and fragrance. In our daily life flowers are a dominating factor for auspicious occasions, expression of colours, beauty, arrangement, wreaths, garlands, fragrance. Apart from this these flowers form a part of delicacies, healthy diet. To name a few in this category are Kachnar, Nasturtium, Calendula, Viola, Roses etc. Lotus flower remains the possession of Lord Brahma and Goddess Lakshmi. Pedestal of Lord Buddha is Lotus. The natural instinct for flowers is practically an unending desire for love, beauty, colours, decoration, fragrance, oils, medicines, garlands, bouquets, wreaths and presentations. It is this demand which makes flower culture a unique economic growth as a perspective of the future industry.

Flowers in Indian customs and tradition has remained an integral part of daily life in one form or the other, right since the start of the civilization in the Indian continent. Cut flower blooms as offering to the deities are sanctified as worship offers in homes and temples and as garlands and extraction of natural colours, oils. Usage in Ayurvedic medicines formed a part of traditional Indian medicine and floriculture. Flowers formed a definite part of grace and decoration in Indian festive ceremonies like marriages, worships, religious congregations, decoration as head gear too. In the far Eastern and Eastern states of India Tamil Nadu, Kerala, Karnataka, Andhra Pradesh, West Bengal flower head gears is a culture and a sign of respect and prosperity. Main flowers of culture were the Tagetes / Marigold, wild Roses, Jasmin. Offerings were not restricted to cultivated flowers but also to the wild one also. Today the flowers have become an essential part of agriculture as an economy generating culture and found its first commercial start from Netherlands in the 16th century. Dutch are the leading technologists of flower culture in the world both in the open and under covers too. Dutch carried alongwith them this tradition to India, but Indian farmer did not follow this culture then. Till Independence, India only confined to the olden customs and traditions of Floriculture of the known flowers Tagetes/ Marigold, Roses and Jasmin in the open system. Floriculture of these species remained confined only to the rural areas very much close to the Indian metros. Usage of flowers now in the form of cut flowers, garlands, bouquets, flower arrangement, oil extraction, drying of flower petals for natural colours, decoration, gift presentation has come up with the growing demand and virtual increase in the number of species and also the time of culture during the year whether Annuals or Perennials.

Present day floriculture has taken the shape of an industry and is the result of centuries of evolution from nearly non existant culture to flower trade. Similar is the pattern which took place in Netherlands in the sixteenth century in the year 1594 when first Green House was established for covered crop culture. Floriculture unlike others is dependent on production and trade facilities available in a country or region.

Birth of floriculture industry in the Netherlands is attributed to the combined effect of its being biggest sea power and higher living standard of Dutch people then in the 16th century and their elegant taste for flowers and floral products. Coming of Dutch to West India and the Dutch East India Company in the early 16th century, only Lilacs were cultured then in the green houses but over the period now almost all the flowers and ornamental plants are produced there. Major flower growing areas in Netherlands around Alsmeer in the western part under protected system of flower culture, spread ultimately to all parts of the world. Along with this, the unique feature which has made Holland as the worlds highest developed system in floriculture is its unique selling system, interest of the farmers for latest technologies in the cultural field of crops. With the increasing culture costs due to labour, heating costs, quality maintenance, technology development, post harvest handling, Dutch had been looking for alternative resources for flower culture in other countries and reducing production costs. Labour, energy are the two components of this intensive crop culture and are much cheaper in India. With this thinking and direction and the growing demand for flowers, ornamental plants and floricultural products the world over, a revolution like thinking thus has been generated in India too. India's share in the floriculture trade as of 1989 was (3%. During the decade of nineties when India had very little to export, people started looking floriculture industry as another innovative dollar earning economy after India recognized Israel as a country in the year 1991.

In the year 1981-82 export earning from flowers was only Rs 84 lakhs and in 1990-91 Rs 7.38 crores which went up to Rs 11.39 during the year 1991-92. Present export for the year 2000-01 is Rs 123.12 crores (APEDA-2002). Status of floriculture in India thus reveals, it has graduated from domestic agriculture to Floricuture industry. Very many major companies and the farmer organizations are expanding and seeking out markets in Europe, Far East and are also complying with various Environmental specifications to remain very much in the world competition and in the Export markets of Europe and South East. Major hurdle in developing trade in the European Union bound floriculture export of cut flowers and other products from India is open air floriculture of the age old traditions, the Duty Tax. India's major competitors in the trade and culture of flowers are from Africa like Kenya, Colombia, Zimbabwe etc. and Central and South American countries who enjoy Duty exemption, where as India must pay higher Duty Tax and higher freight costs as well and then compete (Floraculture International - Nov. 2001).

FLORICULTURE TODAY

Ornamental horticulture of today is not a past time hobby but has commercialized itself through advanced cultural techniques and growing flower demands in the domestic and in the export markets. Floriculture under the open and also under covers in the Temperature controlled Green Houses has come to stay as a commercial viable enterprise. Neighboring countries of Thailand, Malaysia, Indonesia, Singapore, Srilanka are earning a sizeable chunk of foreign exchange through floriculture. Total world trade for Cut Flowers in 1990 was 2.9 Billion US$. India in comparison with its 34000 Ha under Floriculture earned only Rs. 84 Lakhs (0.4 Million US$) in 1981-82 and Rs. 850 Million in 1992-93. Israel which is only 1/3rd the size of Himachal Pradesh is a world leader in Floriculture Technology with 4% share of the world export trade in Cut Flowers up till 1999, innovative technology in floriculture development after Holland, now the position has been taken over by Kenya.

ECONOMIC SURVEY 1994-95

1. Viewing capital formation decline to Rs 10650 Million in 1992-93 in comparison to the year 1980-81 with Rs. 17960 Million, survey suggested need to strengthen the economy by giving incentives for attracting more private investment into agriculture.
2. With its rich agricultural resources and achieving world production standings, contribution of agricultural sector is very poor. Export earning of 133.4 Million US$ during 1993-94 was highly low in comparison to manufactured goods sector commanding a trade share of 75.5% with an export earning of 18789 Million US$ which is 4 times more.

SCOPE OF FLORICULTURE INDUSTRY IN INDIA

Commercial floriculture has become the most profitable agro-industry of today in many developed and developing countries of the world. In the recent years very many developing countries of the world Colombia, Kenya, Zambia, Zimbabwe, Mexico has emerged as potential centers for floriculture crops. Traditional cash crops in Fruits and vegetables are no match to floriculture export and its yearly growing demand, very many countries and growers of the world adopting this as one of their economy bets of the future. Taking a cue from the Israel floriculture export trade with its small holdings and its expertise TATA, ESSAR, RPG, SPIC has diversified to the floriculture trade. India till today has not exploited this venture commercially though with a very good area under flowers much more than the leaders in the world floriculture trade, but this all is in the open Air culture. Ornamental crop culture, also includes in it floricultural products like potted plants, foliage plants. India has the potential of producing both temperate, tropical foliage plants both in the "Cut and Potted plant culture" but that is also limited to domestic demands only. The other fields of imminence are Seeds and Bulbs which at present though being produced are for only domestic culture and do not meet the world quality requirements for colour,

No. of flowers, size, disease resistance etc. which is the consequence of meager or no research direction in this field in India. Export oriented viable floriculture trade requires:—

- Cultivation of flowers, foliage plants much in demand in the export markets of the world.
- Multiplication and culture of only those varieties which are accepted in these markets.
- Production of flowers only at a time when these are much in demand with higher prices in the importing countries.
- Presentation and shelf life increase processes and procedures be adopted, where in the quality is best retained till the destination center and the auction with also increased Vase life.
- Development of infrastructure like cold stores, pre-cool chambers, Reefer Vans, and cold store development at the cargo centers at the Indian Airports and completion as early as possible.
- Development of allied industries like Green House manufacture, Polythene covers, shade nets, Irrigation, Fertigation systems, Humidification and climatic controls and introduction of digital technology unlike the other developed countries of the world.
- Provision of direct Cargo Air flights to the auction centers of Holland, Germany, United Kingdom. Floriculture under European conditions is becoming very professional and commercialized with the controlled culture and that too mechanized, because of scarcity and costly labour, non availability of skilled personnel. Production costs are rising on other counts too i.e. Electric charges, costly Input costs. In India both skilled and un-skilled labour are available in abundance and is very cheap. Input costs are also very cheap.

In view of all this there exists a tremendous potential under Indian conditions with varied agro-climates, availability of human resources both skilled, un-skilled with cheap electricity and the will of the farmer for determined entry into this lucrative trade. Floriculture as it should be viewed is the branch of Ornamental Horticulture and is not merely confined to cut flowers, ornamental plants for pleasure but has taken the shape of an economic enterprise. Floriculture as a commercial venture being the most profitable industry encompasses in itself the trade of Cut flowers, Ornamental foliage plants, Potted plants, Flower bouquets, flower arrangements, other floriculture products. Due to very little thrust and no proliferation of latest technologies in floriculture into action oriented programs, farmer and the corporate remained ignorant of the best developing economy. Annual growth rate in the floriculture sector is 10–15% and the developed countries are focusing their production off shores in other countries. Migration of Dutch to other countries of South America is the live example of this trend. American Flower Industry predicted the demand for cut flowers and ornamental plants will continue to exceed production as the market for these products will expand at a faster rate. According to trade statistics, total world imports of cut flowers amounted to $ 3716.8 million in 1995, up from $ 3351.7 million in 1991, an increase of nearly 11% during the five year period. During the past three decades several developing countries like Kenya, Columbia, Mexico has emerged as potential production centers for flower crops and contribute 6% of the world production. Floriculture is the depending economy of very many countries and Holland is one amongst them. Israel, Thailand, Malaysia, Indonesia, Srilanka, Singapore are earning a sizeable part of their foreign exchange from floriculture. Viewing the achievements of these small countries in comparison to India, we must evaluate our potentials of facing challenges of competition, trade study for market development, product identification and the areas to concentrate for culture in India, latest technological developments in Post harvest and development of infrastructure etc.

In floriculture, export can be viewed for cut flowers, potted plants (foliage and flowering), cut foliage, seeds and other planting materials. With the growing demand of floriculture the world over and the rising production costs elsewhere, Indian production is very cheap because of cheap labour, electricity and input costs. India thus has the capability to produce flowers which are in great demand, only the accepted varieties in those countries. Indian production of flowers and its products to the world markets has another capability of phasing its production when production in the importing countries is lowest. Winter months right from October till April is the period when production in those countries lowest and market is too responsive for its imports from other world countries.

For maintenance of product quality, post harvest technology is the other area which is of utmost requirement. India has entered in this to help maintain product quality and has opened its Temperature controlled Cargo complexes at Delhi, Mumbai, Bangalore where it receives the consignments to be air freighted to the trading markets in the world.

Post harvest chemicals like hydration and holding solutions are being imported into India from Holland for increasing the vase life and keeping cut flowers fresh.

International Trade Centre UNCTAD/WTO Geneva in the year 1987 published its survey report on floriculture showing the major world markets in trade of flowers and plants only to provide advice for establishing an export oriented floriculture industry. These markets are France, Germany, Netherlands, Switzerland, United Kingdom, Canada and the United States of America. Major cut flowers of trade which are in demand in order of volume of trade are Roses, Carnations, Orchids, Gladioli, Chrysanthemum and others. Report further describes the market requirements and the type of products in demand, basic understanding of market structures, competition, prices, market access, import regulations, customs duties, importers addresses. This study is a background material for the Government bodies, growers, exporters, International organizations, Trade Associations.

Major emphasis in buying the flowers and other floral products is on quality, post harvest loss reduction technologies such as harvest maturity, ethylene control, packaging technology, controlled and modified atmosphere storage, use of bud opening and floral preservative solutions, establishment of grades and standards for different floral products are the needed inputs.

Floriculture has been identified as the thrust areas of growth in export earnings with an annual growth of 25-30%. India's central location on the world map in comparison to most important consumption centers in EU, SE Asian countries affords good opportunities. Cheap labour with skilled manpower and electricity availability and inflation under control have created congenial conditions for industry set up. Government has realized though belatedly, but has encouraged domestic producers by providing subsidies, grants, low interest loans and also allocated funds for research is a definite direction and creation of different bodies to deal with this development and requirement of the country and industry as a whole. Government support though is very meager in comparison to the developed economies but is a positive sign of its determination in upgrading the system and to bring it at par with the developed countries of the world in this trade.

FLORICULTURE : NON-DEVELOPMENT IN INDIA

India is a country with potentials is known to every body and the Dutch too in the world. With 75% population dependent on agriculture in India and also growers carrying on the heritage culture of flowers in the towns, villages around the metros were too curious to enter the export markets in floriculture, but every time had to remain contented with domestic marketing only. India with 29 Agriculture Universities, 6 Research Institutes and very many other stations and centers involved in the development of floriculture in the country did not direct the energies for farmer or market oriented research. As such area under Floriculture though 70,000 Ha is all an open culture issue. Floriculture for commercial trade and International marketing requires well set up controlled culture facilities and also post harvest handling of this delicate product.

Commercial Flower culture was hardly a non business proposition prior to 1990, though Indian floriculture Market is worth 400 million US$ at retail and in International quality flowers about 100 million US$. Floriculture exports are only at 30 million US$. For high quality cut flower demand in the Indian metros local markets have been catering to over 70% of the requirement. Foreign buyers and distributors who so ever visited India prior to 1990 discouraged Indian grower, fully knowing about his potential, skills and becoming a competitor for them in the Dutch auction markets. Central and State Governments working seriously in tandem created infrastructure for flower handling and other requirements by investment opportunities for the FDI and local corporate bodies and growers and also setting up of 3 cold stores and 2 whole sale auction centers. This effort was right since 1996 being made possible by sincere and dedicated efforts. However realizing fully well the direction of the Indian Government and expanding cut flower and potted plant business from other continents forced these big Dutch players to enter the Indian market and with this they entered into Agreements only for the delivery of short term gains through Inputs, short term interests, in the form of contracts, Fee's, Commissions on sales, Advisory, Consultancy services. Even with this very few projects were set up by Dutch in India.

Credit here must be given to ISRAEL who gave a real tough lesson to these Dutch bargainers by competing in every line of Input supplies, services, consultancy etc in floriculture and India stood of its own after 1991. Israel went further in growing systems and providing demonstration centers arranging collaborations with local Governments in actual Farm Projects beyond Floriculture and training Indian Farmers and also making them to participate in annual Fairs, Seminars, Conferences which was never so before. Israeli irrigation, Green House fabricating companies set up joint ventures in India for marketing Drip irrigation and Green Houses in India. It was only then the need for Cargo system development was felt urgently for the export of cut flowers from India to the destination markets in Europe and other countries (Floraculture International, November, 2000 by Sid Mukherji, MD Sun blossom Florals).

<u>CHAPTER - 3</u>

FLORICULTURE AS EXPORT TRADE

EXPORT SCENARIO

Very many countries of the world have built up their economies on agriculture and examples to mention are – HOLLAND on Floriculture, NEWZEALAND on Kiwifruit, ISRAEL on Agriculture and Horticulture. In the African continent the LDC (Least developed countries) like Burkina-Fasso, Ivory Coast, Cameroon are not far behind these developed economies to have their share from Hort-agro products. India's export share in the world trade for export has always remained very dismal and floriculture in particular. In the year 1990-91 Horticulture products commanded a share of 18.42% in the total agriculture exports during the year for Rs 6012 crores. Low performance in export of agricultural products from India is attributed to a variety of reasons in which major one's are :—

- Lack of Infrastructure like Roads, Cold Stores, Refrigerated carriers.
- Little or no knowledge on flower culture and floriculture products much in demand in the export markets of the world.
- Un-awareness of Post Harvest Management Technology.
- Lack of extension follow up of Post Harvest systems.
- Non compatible germ plasm.
- Export market Intelligence Availability of Cargo services and high freight rate.
- Research efforts of the Agriculture Universities of the country is not directed fully for floriculture.

- Research programs taken up are not as per practical requirements but for academic purpose studies and the staff is not fully employed for this sector.

EXPORT POTENTIAL

India's export potential lies in its central location in the routes of world trade centers, proximity to major consuming centers of European Union, South East, Gulf countries considering the Air and Sea routes. Desired strategy has been evolved by the Government of India through its concerned Ministeries and Departments of Agriculture, Commerce through National Horticulture Board, APEDA and International Trade Promotion Organization to work in tandem to remove all impediments and constraints in Export procedures, Import of Technology, Exemption of goods from Customs and Excise duties, concessional Air Freight, Liberal loan financing and to make its products highly value added and making more competitive in the world markets to gain an access to the Floriculture growth of 15% per annum. Government of India on the recommendations of the expert committee, projected the agricultural scenario as growth oriented and declared incentives as under for 100% Export Oriented Units apart from others which already exist in the financial and Technical know how side.

Export Boosting Government Incentives

1. Zero Import Duty : Seeds, Bulbs, Tubers, Cuttings or Saplings for sowing / planting have been exempted from the levy of Customs duty on the import of these items.

2. Plant Quarantine : Clearance within 6 hours on the import of seeds, plants, tubers, bulbs.

3. Certificate of Origin : Plant Quarantine and other Officers can also issue the certificate of origin in respect of Horticultural exports.

4. Import under OGL : Flowers, seeds, Tissue culture material of any plant origin is allowed for import without the need of an Import permit.

5. Cut on Import Duties : Import duty on seeds, Development machinery, soil preparation and specified goods of Green House have been brought down to 25%.

6. 50% Domestic Sales for EOU : Units engaged in floriculture can avail of the benefit of duty free imports under the EOU / EPZ scheme even if they export 50% of their production.

7. Infrastructure Development : Cold stores for the export of floriculture consignments awaiting CARGO at the International Airports have been developed at Delhi, Bombay, Bangalore. Cold stores at Madras and Trivandrum are coming up very shortly.

8. Subsidy on Air Freight: For the export of Cut flowers and Tissue culture plants by Air a subsidy on Air Freight has been allowed by the Government. This is subject to a maximum of 25% of the IATA freight rates or Rs 10 per kg. for Europe and USA and Rs 6 per kg. for export to West Asia and South East Asia.

9. Custom Bonding : 100% EOU's in some of the agriculture related activities including floriculture have been granted exemption from the requirement of Custom Bonding with permission for Duty free import of specified goods being used in production/ manufacture or packing.

10. Import of Specified Goods : Import of specified goods for Green Houses allowed against the concessional duty of 25% advalorem, subject to obtaining certificate from the Ministry of Commerce to the effect that goods are required for Green House. The allowed goods are UV retardent Fibre Glass / Plastic sheets, Low draft horizontal axial Fans with energy efficient motors, Fogging machines, Automatic shading system, Drip Irrigation system, Humidifiers, Pads for cooling system, Heat exchangers for steam sterilization of soil, Flower grading machines.

The Project design thus involves in the Flower culture under covered conditions with all the controls on Climate, weather and soil, moisture, Light with a perfect control on production, harvesting and post harvest management techniques with a provision of a Green House, Pre-cool chamber, Cold store, Package and Grading hall, Reefer van for transport to the Cargo center thus keeping in tact the cold chain process of the post harvest management.

Opportunities : Lot of potential and tremendous opportunities exist for the development of floriculture in India because of availability of both skilled and un-skilled labour at very cheaper rates in comparison to European and American countries. Along with this the cheap inputs like electricity energy, good water supply. Viewing this the annual growth rate of floriculture has been put at 10-15%. Developed countries are focusing their production off shores and in potential countries which can deliver the results for their faith. Developing countries of the world have only 6% share in the export trade of floriculture. Trend of the global markets is very encouraging in view of the encouragement being given to the new entrants in the trade. Singapore is opening its auction market. Above all is the will of the Indian grower to learn and adopt new technologies, which are very high return paying ventures.

Top Ten Floriculture Products of World

FLOWERS		POT PLANTS	
Rose	Freesia	Ficus	Chrysanthemun
Chrysanthemum	Gerbera	Dracaena	Yucca
Carnation	Gypsophylla	Kalanchoe	Saintpaulia
Saintpaulia	Alstroemeria	Rhododendron	Hedera
Tulip	Cymbidium	Begonia	Poinsettia
Lily			

FLORICULTURE EXPORT PRODUCTS

1. **Cut Flowers**
   - Roses
   - Carnation
   - Chrysanthemum
   - Orchids
   - Gladiolus
   - Tulip
2. **Minor Cut Flowers**
   - Gerbera
   - Anthurium
   - Caladium
   - Lilies spp.
   - Freesias
   - Gypsophylla
   - Iris
   - Narcissus
   - Statice

3. Potted Flowering Plants

- Chrysanthemum
- Poinsettia
- African Violet
- Begonia
- Geranium
- Pitcher plant
- Easter Lily

4. Foliage plants

- Aglaonema
- Palms (King Sago, Areca palm, Parlor palm)
- Dieffenbachia
- Ficus spp.
- Ferns (*Pteris* and *Adiantum* spp.)
- Araucaria

Floriculture industry comprises of cut flowers both major and minor, potted plants, foliage plants. In this whole trade of export major portion of 70-75% is covered by cut flowers. Total sales of the 12 major export floriculture products in the Netherlands auction center during the year 2001 amounted to a turnover of 1462 million Euro's. World cut flower area in the year 1985 and 1989 was 48000 and 56000 ha respectively. In this 13-28% area was under Green House culture. Major countries in the culture of flowers Netherlands, USA, Italy, Mexico each contribute nearly 10% of the area for this (Floraculture International).

Table. Showing Value and Quantum of Sale during 2001 at Aalsmeer Auction Center Netherlands

Flower	No. of stems sold in millions	Sale Value in Euro	Price / stem in Euro
Roses	3200	653	0.27
Chrysanthemum	1260	254	0.15
Tulips	1100	177	0.16
Lilium-Oriental	132	76	0.58
Lilium-Asiatic	143	38	

Gerbera	43	21	0.21
Cymbidium	10	33	3.26
Freesia	291	62	0.15 (Single) 0.16 (Double)
Dianthus	480 (Spray-239) (Standard-241)	58	0.12
Alstroemeria	277	42	0.15
Anthurium	60	37	
Lisianthus	91	11	0.27

ORCHIDS AS DOLLAR EARNERS FOR INDIA

India is a big home for Orchid culture, being a center of origin and possessing a very wide diversity in the species and areas of Orchid culture in the states of Arunachal, Nagaland, Mizoram, Tripura, Meghalaya, Assam, Tamilnadu, Karnataka, Himachal, Jammu and Kashmir, Uttaranchal. Commercial culture for Orchid cut flowers and potted Orchids is yet to take a shape in these areas, because of no extension and awareness exercise ever taken to expose the farmers of underlying wealth in this flower. Our very near next door neighbours are Thailand, Malaysia the largest world Orchid producers, exporters followed by Singapore and New Zealand. In Europe Italy is the only country to take a lead in Orchid culture. Japan is the biggest importer of the Orchid cut flowers and potted plants. In this total trade of Orchids share of the potted Orchids is only 15% and the rest is for the cut flowers. India with its vast resources, potential of manpower and the interested farmer, advanced education availability, best seasons for year round culture, Tissue culture Laboratories for plant multiplication, governments intuition to enter into the arena is very much behind in this field of floriculture products. Then where lies the exact lacuna of not turning the tables. It is the extension education and awareness. Commerce of Orchids as it was during the year 2000 in the world is as under. There was a total trade of Export/Import to the tune of 150 million US$, which had a share of 128 million US$ for cut flowers and the rest for potted Orchid plants. In this trade Asia dominates the scene of trade.

Export / Import of Orchids During 2000

Exporting country	Product Type		Total Trade Volume Mill. US$
	Cut flowers	Potted	
Thailand	50.0	4.000	54.000
Singapore	7.7	0.008	7.708
Malaysia	2.8	0.015	2.815
New Zealand	0.830	0.093	0.923
Italy	0.652	0.399	1.051

Japan is the world's largest cut Orchid import market, where cut Orchids is a passion in a house, flower sale shop. During the year 2001, total sale of cut orchids by the Japanese auctions was to the tune of 230 million US$. This total trade is inclusive of the own Japanese production too, which specialize in Dendrobiums. During the year 2001 Japan imported 42% of all the cut flower Orchids traded around the world, which constituted 50% from Thailand, 19% from New Zealand, 13% from Singapore and the remainder from Taiwan.

CHAPTER - 4

INDIAN ECONOMY AND HORTICULTURE

INTRODUCTION

Indian economy is basically a Farm oriented economy where 75% of our masses live in rural areas and are dependant on their small holdings. Sluggish growth of economy is attributed to non-symbiotic relationship between the agriculture and the Industry. This is evidenced from the latest survey report of 1994-95 where in gross investment remain stagnated at Rs. 4617 crores during 1992-93 over the year 1980-81 when it was Rs 4636 crores. Second policy factor is that any of the agriculturist/farmer who grows or cultivates a particular product at his farm land has no legal sanction to convert his product into a value added component. Indian Farmer is poor and illiterate, un-aware, un-exposed, but well understands latest technologies, corporate planning and Management which is the soul of the Industry. These are the major facets of industrial agriculture. On the technological and cultural side other facets of agriculture are Post Harvest Management, Floriculture, Food Processing, Fisheries, Bee-culture, which can improve the economy of the Farmer and the nation. With the entry of the Corporate sector in the Fruit Processing, buoyancy of investment has increased. Corporate sector with every access to Financial arrangements, planning, technology and management, market research and analysis, Information technology etc. enjoys all these approach's at his disposal in comparison to the Indian farmer. Direction was provided by the Government and the various Industrial bodies to work in harmonious attitude with the farmer thus to improve upon the dis-integrating and staggering

Indian economy. APEDA 's thrust to value addition through Cold chain and recent seminar of the CII, opening up of Israel model type Floriculture trading, are if some indicators, then this is a definite direction for agriculture sector after the green and the white revolutions.

Horticulture thus has imminent role to play in boosting the economy through Fruit, Vegetable, floriculture, Processed Food segments. Floriculture and Floricultural products available in India possess the potential of best culture through varied agro-climatic conditions, energetic farmers and highly educated and skilled Human resources for culture of any flower in the world floriculture trade. With this thinking Government conducted an Export Group survey on Floriculture in India. Recommendations of this group are :

- Varied agro-climatic conditions provide enormous scope for cultivation of horticultural / floricultural crops.
- Areas around Bangalore, Trivandrum, Pune, Nasik and Delhi be developed for Hi-Tech Floriculture Industry.

IDENTIFIED FLORICULTURE ZONES OF INDIA

India possesses every type of climate that exists in any part of the world countries is a unique phenomenon for this country. However every such area can not be recommended for floriculture, as heavy rainfall, high humidity, high and too cold temperatures are the extremes which can be suited to a particular type of flower culture in these areas. Thus high rainfall and humid areas of the South and the North east are suited for Orchids and the mild temperature areas for Roses, Carnations and Chrysanthemums. As such whole country has been divided on the basis of climatological, soil characteristics in 5 different zones.

1. **Northern Plains** : Areas encompassed by Ganges, Yamuna rivers characterized by alluvial soils comprise the states of Rajasthan, Punjab, Haryana and Uttar Pradesh.
2. **Eastern Gangetic Plains :** Areas comprising the states of Bihar, Jharkhand, Chhattisgarh, West Bengal, Orissa fall in this.
3. **North Eastern Region :** Plains and hill areas of seven sisters i.e. Assam, Meghalaya, Arunachal Pradesh, Nagaland, Manipur, Tripura, Mizoram and now Sikkim

too form this great part in this region. Heavy rainfall and high humidity are the peculiar features in this region. These regional states are in very great proximity to soil, temperature and climatic conditions of Thailand, which has a name in the world for its Orchid culture. These states are very rich in biological diversity and have origin for record number of Orchid species in the world.

4. **Central India :** This region with arid climate, alluvial red soils comprise the states of Madhya Pradesh, Gujarat, Maharashtra.
5. **Deccan Plateau :** States of Karnataka, Andhra Pradesh, Kerala, Tamilnadu, Pondicherry are covered. However, areas like Hyderabad, Nasik, Pune, Bangalore has attracted much attention of the floriculturists, because of very moderate climate, low rainfall and deep soil.

Table: Showing Indian States identified for different flower groups

S.No.	Identified Zone	Flower Groups
1.	Delhi, U.P., Punjab	Rose, Carnation, Chrysanthemum, Gladiolus
2.	Karnataka	Rose, Carnation, Chrysanthemum, Gladiolus, Ornamental Foliage Plants, Seeds
3.	Kerala	Orchids, Anthuriums, Foliage Plants
4	Maharashtra	Rose, Carnation, Gladiolus, Dahlia, Chrysanthemum, Aster, Tuberose, Indoor Plants
5.	North Eastern States	Orchids, Amaryllis, Gladiolus, Lilium, Gerbera, Salvia, Anthurium, Cactii, Foliage Plants
6.	West Bengal	Lotus, Tuberose, Jasmins, Dahlia, Chrysanthemum
7.	Jammu & Kashmir	Gladiolus, Lilium, Carnation, Rose
8.	Himachal Pradesh	Gladiolus, Bulbous Plants, Seeds
9.	Southern Hills	Rose, Jasmin, Tuberose, Carnation, Orchids, Chrysanthemum

Leading flower producing states of India are Tamilnadu, Karnataka, West Bengal, Andhra Pradesh and Rajasthan which produce flowers in the open and green house systems. Percentage of area under production in each state during the year 1989 is as—

State	% of total area in India
Tamilnadu	24.7
Karnataka	23.1
West Bengal	9.3
Andhra Pradesh	9.0
Rajasthan	7.4
Others	26.5

(**Source** : Biotech Consortium India Ltd.- Floriculture Update)

Thailand on the Eastern border of India and just next door to the seven sisters is the largest exporter of Orchids. Arunachal Pradesh, Meghalaya are a home of Orchids and also possess similar conditions as Thailand. Another exotic plant from Meghalaya which stands the export potential is the Pitcher plant both as a cut flower and potted plant. Thus both these orchid and Pitcher plant can be made as a gold mine for these areas, which can be further added to the flower producing states of India. Very good research work is being undertaken by Meghalaya University and extension work is also in the primary stage, which needs concerted and dedicated attention from the Government for immediate results.

CHAPTER - 5

FLOWER CULTURE UNDER COVERS

COVERED CROP CULTURE

At one time farming was looked as a vocation, with profit margins looked or taken care by the Government through subsidies and other support measures. But now with the intervention of WTO in globalising the trade and bringing every country at par with the developed countries, subsides has been slashed. In view of all this, any business which can not maximize its profits and produce, a suitable return on capital can not justify its continued support. Farmer has thus to take its own decision in raising profits which can be through higher productivity, reduction in cultural costs, reduction of post harvest losses and bringing up the product quality to compete with the best in the market. Covered crop culture has all the capabilities to deliver the desired results and ensuring proper return on capitals.

Necessity is the mother of invention and innovation of crop culture under covers, was started long back by the Romans. Romans and the Greeks in the West were in the fore front of civilization during that period in the West. Romans desired to have prolonged crop periods and availability of vegetable crops even during winters when there is enough of snow and to protect their crop plants from the vagaries of weather also. It was only then that Romans developed a system to culture crops under protective covers. The system provides potential to modify environment i.e. temperature, Light, Humidity, water and air movement, conditions to better manage pests and diseases. Generic term referring to the use of a transparent or partially transparent material supported by a structure to enclose an area for cultivation of crops, propagation of plants came to be known as Glass House. With the passage of time and addition of innovations and technology into covered crop culture this has lead to the manufacture of best of the Green Houses which provide the desired climate and environmental controls in the Green House for the best plant growth and yields. Covered crop culture is done in the Glass and Poly Houses. These are the prominent components of this industry.

Covered crop culture is the tradition to raise early crops, protect from winds, frost, cold and heat. This has been necessitated with the consumer demands, gaining an access to early market, protecting crops from fluctuating climatic conditions and saving from pest menace. Be it nursery raising of vegetables or vegetable culture or flowers it used the technology of covers for getting the early benefits of the market, crop protection etc. with full protection from climatic vagaries. Initially it was the Glass Houses but now the culture has modernized with the usage of Polythene film, which works as a cover. Green houses are thus either Glass or UV Polythene or Tunnels or Cloches which provide plant growth under covers. Green houses offer a higher return per unit area than the field based production of the same crop. Intensive cropping has thus been generated as a viable business opportunity on small and large areas. A sophisticated Green House can produce 15-20 times more produce than a field of the same area. Protected crop also offers substantial environmental benefits in management of water and pesticides.

Dutch are widely recognized as the matriarch of Green House industry in the world. Approximate area of production in Green Houses in Netherlands is 12000 ha. Maximum area of 7000 ha is in Westlands Aalsmeer District and next to this is the Naldjwick District with 3000 ha. main crops cultured in these are Roses, Tulips, Chrysanthemums, gerbera, Carnations in Flowers and Tomato, Capsicum, Eggplant, Cucumber in vegetables. Minimum viable unit under the Green House is 4000 sq. mtrs in Netherlands.

Advantages of Green / Poly House

- Provide an optimum growing requirement.
- Create longer growing seasons.
- Grow crops out of season.
- Faster growth and higher yields.

- Protects crops from cold, hot and fluctuating weathers, hails, damaging storms and dry winds.
- Control on pest and diseases.
- Protection of sensitive crop plants from intense sunlight.

Green or Poly House is thus a composite unit of structures to stand, covers and auto-controls where in plant culture is done as to very specific for every individual plant as per total requirements of light, water, nutrients, other atmospheric environmental requirements.

Green Houses : of today are pre-fabricated structures of steel and UV polythene film with a full climatic and cultural controls contained in these available in the market with pre-defined designs. These green houses has inbuilt design specifications and variations of span, width, height, UV protection film, Vent openings, irrigation and fertilizer controls. Temperature controls are either pad and fan cooling or from Gas cooling units to maintain the desired temperatures in the Green House in the tropical countries and it is otherwise to provide heating in the temperate region countries. Irrigation to the plants has been so improvised that each plant gets the required amount of water through the drip system and fertilizers application with irrigation water. In the green houses floriculture has attained all types of sophistication in cultural management systems so that the quality and the shelf life of flowers is not only increased but production status is also kept at its highest level. Best usage of covered crop culture has been made in the raising of cut flowers, foliage plants, potted and cut foliage plants. Potential in this culture lies in highest returns and capability of taking risks, which otherwise is not there in other crops. For crops like Tomato's, Cucumber's, Melons, raising of nursery etc. are undertaken in Tunnels and cloches which are much cheaper.

Why A Green House : Before taking up the job of erecting a Green House, it must be understood and known, what for the house is being raised whether for propagation or crop culture. Design and types of Green houses are very different today for each and every specified purpose and available on demand at competitive prices. In order to have best benefit and advantage of the technology, expert opinion must be sought before hand. This exercise saves money, time and provides un-known benefits of technical aspects which otherwise remain ignored.

- Cool / Warm house for propagation.
- Efficient labour saving facility.
- Type of crop culture whether Floriculture or Vegetable or Herb culture.

The other connected aspects to be considered before deciding the construction of the Green House are :–

- What is the desired volume of production to be raised.
- Expected return.
- How much you can afford to invest.
- What is the source of water and whether this is clean.
- Does the system requires heating or cooling.
- Does the system require automation of the operations like irrigation / misting, fertilization, Integrated pest management, environmental controls.
- What type of zone this Green house is located whether hailstorm or windy.

Types of Green / Poly Houses

1. Gable
2. Sawtooth
3. Skillion
4. Tunnel (Igloo)
5. Open roof (Less maintenance, affordable costs, best ventilation)

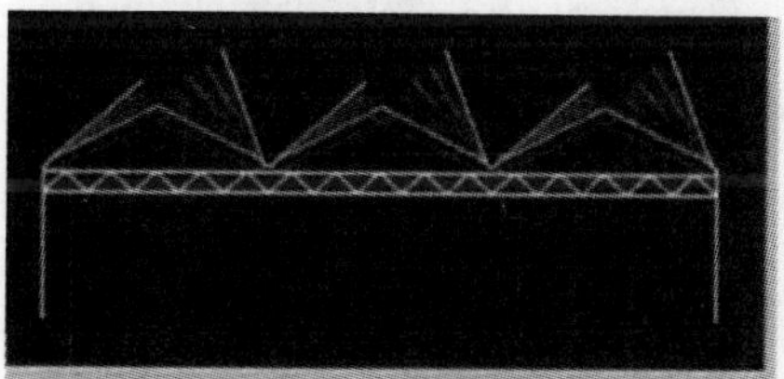

Open Roof Green House

Multispan Green House

Bedding and Potting/Growing Media: Potting media is the best possible substrate which creates most ideal growing conditions for the plant material. In the Green house culture bedding or potting media is prepared as per plant requirement and is a mixture of sand, soil, fibrous compost from biogenic wastes, moist granulated peat, coir peat or cocoa peat, perlite, expanded plastics, vermiculite, pine bark, wood fiber and shavings, bark compost etc. form the bedding or potting media. Potting media differ from agricultural soils in their physical and chemical characteristics. Fertility status of the potting media is a very vital factor. Nutrient levels are well optimized and are crop specific in order to obtain maximum crop growth. Due care is given for the balancing of the nutrients like calcium, potassium and magnesium. Micronutrients are added in the media to ensure a proper balance with a proper care on the pH. Another factor best noted is the moisture holding capacity of the potting media, which determines the root development and growth of the plant. In the developed countries where greater emphasis is on the potting media, all the commercial medias in the market are duly tested for presence of constituent levels. Best potting media which has been tested and found fit for floriculture is Coco-peat a product of coconut coir prepared by decomposing the fiber. Other potting substrates in this category are wood fiber, bark and bark compost. All these materials can be used in combination to form own substrate and potting media. In India it is yet too far off to think on these lines. pH is lowered by the addition of Iron sulphate and is increased by adding ground lime. However, evaluating criteria for the minimum and adequate levels of nutrients is as under :–

50% Vent

Table : Showing Potting Media Evaluation

Parameter	Unit of Measurement	Rating		
		Low	Adequate	High
Soil pH	-	< 5.0	5.0 – 6.8	> 6.8
Conductivity(mature plant)	mmhos/cm	< 0.7	0.7 – 3.5	> 3.5
Conductivity(young plant)	mmhos/cm	< 0.5	0.5– 2.0	> 2.0
Available N(NO_3-N)	ppm	< 40	40 – 200	> 200
Phosphorous	ppm	< 3	3 – 10	> 10
Potassium	ppm	< 60	60 – 250	> 250
Calcium	ppm	< 80	80 – 200	N/A
Magnesium	ppm	< 30	30 – 70	N/A
Sodium	ppm	-	0 – 80	> 80
Sulfur	ppm	< 20	20 – 200	> 200
Boron	ppm	< 0.5	0.7 – 2.5	> 2.5
Iron	ppm	< 15	15 – 40	> 40
Manganese	ppm	< 5	5 – 30	> 30
Zinc	ppm	< 5	5 – 30	> 30
Copper	ppm	< 2	2 – 30	> 30
Chloride	ppm	< 2	<45	> 45
Fluoride	ppm	-	< 5	-

Source: Adapted from Saturated Media Extract Method by D.D. Warncke. NCR Publication No. 221 (1998), pp 61-64

Hydroponics : Alternative to potting media was the innovation and presentation of Hydroponics during the year 1993 in Netherlands. In this, culture is done both in the stagnant and running water's. Running water enables the roots to take more Oxygen. It is easier to apply fertilizers to bulbs in right proportion in the Hydroponics. Duration of rooting period is 6-16 weeks in substrate where as it is 3 weeks in Hydroponics in the beginning of the season when temperatures are low. But when temperatures are warm rooting period comes down to 1 week which continues till the end.

Auto Controls : Application of computeronics in floriculture has not only modernized and automated the management system for air, soil temperature, light intensity and opening of vents, application of irrigation water and fertilizers but has led to quality crop culture where in the defined crop requirements are met in full.

CULTURAL MANAGEMENT

ROSES : Owing to their wide variety in colour, type, fragrance, growth habit and plant size have very widespread consumer acceptance. Roses are Shrubs, Large flowered, Clustered flower types, Polyanthas and miniature plants. In the world trade of cut flowers Roses with a trade share of 39% are at the top. The two type of Roses as cut flowers in demand are the HT's and Floribundas. Temperature has a profound influence in the growth and quality of Roses. Soil temperature is the other vital factor, where 18.3°C is the mean soil temperature has been shown to produce maximum number of quality flowers. Commercial Rose cultivars are best grown at 21.0°C day temperature and 15.5°C night temperature. Levels of Carbon dioxide have also to be monitored, since this has a profound effect on the vegetative growth, yield and quality of Rose flower.

There are distinct differences in the American and European markets, which are consumer appeal based. In the United States Rose imports 95.6% comprise of almost exclusively long stemmed large flowered varieties and in contrast the European 65% demand is for the small flowered varieties. Netherlands was the largest supplier of Rose to European countries with a share of 76.6% in 1991 which came down to 60.2% during the year 1995. Latest developments for the European market is the expansion of Ecuadorian production and development of new Rose industries in India, Malawi, Nigeria, Swaziland and Tanzania. Import of Roses into USA and EU are showing a rising trend. Imports which were of the order of 2261.1 million stems during the year 1991 has risen to 3341 million stems by 1995 in both USA and EU.

Civil Engineering Requirements : Erection of the Green House influences the best of the crop culture, management and yields. Thus it must enjoin in it, all the facts pertaining to the purpose, volume of production, operation automation, light transmission and requirement, useable space, temperature controls, air movement and ventilation etc. All these factors will determine the shape of the Green house and the type of structural materials and the covers to be used. Shape of the Green House structure influences :—

- Amount of sunlight transmission.
- Amount of natural ventilation.
- Useable internal space.
- Condensation run-off.
- Heating / Cooling requirements.
- Efficient use of structural materials.
- Height of the Green House.

I. Green House Coverings : Vital parameters of the Green House are Temperature and light. Both of these are strongly influenced by the type of covering Glass has been the traditional covering, because of its favourable properties, which are the high PAR (photosynthetically active radiation) transmission, good heat retention at night, low transmission of UV light, durability and maintenance costs. The biggest drawback of Glass is the initial cost as compared to other covering materials. Generally the coverings used are Glass, Plastic sheet, Plastic films and Protective covers.

- **Plastic Sheets :** There are 3 types of plastic sheets, which are polycarbonate, acrylic (polymethylmethacrylate) and fiberglass. Plastic sheeting products are more durable than plastic films and have fairly good heat retention, good initial transmission in the PAR and low UV light transmission.
- **Plastic Films :** This is the most common and cheap material. The type of films available are Polyethylene, EVA (Ethyl vinyl acetate) and PVC (poly vinyl chloride).

Additives are added to these films to make them to the desired standards for light transmissions, heat loss, capacity to reduce droplet formation, transmission of particular wave lengths of light, capacity to reduce the amount of dust sticking to the film. These additives are :–

1. UV absorbers and stabilizers.
2. Infrared absorbers, reduce long wave radiations, minimize heat loss.
3. Long wave radiation (2500 – 40000 nm) absorbers reduce the loss of heat radiated from the materials and objects (plants).

4. Light diffusers.
5. Surfactants.
6. Antistatic agents.

- **Other Additives**
  1. Colour pigments
  2. Fluorescence.
  3. Glossy surfaces, may repel insects.

- **Protective Coverings :** These coverings are apart from the usual covering of the Green House. These are for shading, thermal screens and screens for pests.
  1. Shade cloth
  2. Solar and Thermal screens
  3. Insect proof screens.

II. Height : For creating larger volume of circulation, with stable environment which is better for the crop, in the Green House, height of the Green House to be kept is of prime importance. This may result in higher costs of materials and increased cooling or heating expenses.

III. Ventilation : Heat develops inside the green house and it is very important for good management and higher crop returns to maintain optimum temperature conditions and also air composition of the desired levels. Thus ventilation is another important feature of the Green House.

IV. Light : Plant growth especially in the covered structures is much affected and influenced by both the quality and the intensity of light reaching the leaves. Light energy in the plant is the vital process and is used in the process known as photosynthesis to manufacture Carbohydrates. These carbohydrates are then used to produce other compounds needed by the plant. Respiration is the process in which the trapped solar energy is later released. Amount of the light entering a Green House is influenced by the :–

1. Orientation of the structure.
2. Materials used in construction and covers.
3. Shape of the roof.

Green house should be positioned North-South to provide more uniform light and reduce the shading effect of the support structure.

CUT FLOWER FLORICULTURE

Project : Green House Rose culture under controlled atmosphere

Area : 5 Acres
or
2 Ha

For covered crop commercial Rose culture minimum area has been reckoned as 2 Ha or 5 acres for cut flowers. Project cost is divided under Major and Minor heads right since the project is conceived and production starts. Heads in the Cut Rose Flower project are :—

1. Land Cost and Development : This includes the Land purchase cost, leveling, layout of area, Roads and metalling of Roads, development of water and its resources for irrigation through Tubewell sinking.

2. Plant and Machinery : Covered Rose culture requires the setting and erection of the Green House complete with automatic Atmospheric irrigation, fertilizer application, light controls. This complete structure requires Polyethylene Film for Green house cover, Insect proof net, Electric curtains, irrigation system with Drippers, sprinklers – system controlled by filters for the filtration of sand and anti-calcium treatment of water connected to Fertigation system. Climate and environment control is done by the electric fans and cooling pads, air mixers, each regulated to control moisture and soil humidity, temperature with the help of sensers. Plant protection is taken care by the insect killers and sulphur evaporators. Green House automation is further controlled by the computers leaving no chance to human errors in the system for Climate, Environment, Humidity etc. Computers form a part of the Plant and Machinery of this system. Live Rose plants and tools for pruning and cutting of Roses also form a part of this head.

3. Civil Structures : This includes Administrative Block, Cold Store, Cool chamber, flower receiving and grading shed, Generator room, computer room, Filteration room, Open water tank, Green House Gutters and Drainage system.

4. Duties, Fees, Contingencies : Custom duties for the import of plants and machinery, material shipment costs, transport from Landing port to the Project site, purchase of furniture,

Buckets, trollies etc., Fee's for technical Know-How purchase from the executors of the Project, Engineering planning, Erection Fee's fall in this head. Rose Plants of different varietal mix - 1,30,000 in number will be planted in all the 4 unit of the Green House with a density of 7 plants per sq. Mtr.

5. Working Capital : Expenses on account of fertilizer purchase for 3 months, Air Freight for cut flowers for one month, packing material i.e. Fiber Board cartons, Polybags, Rubber bands etc. for one month, Staff salary and wages for one month for this head.

6. Utilities : Regular Electric power supply, Water supply, Heat generation from coal / diesel for all the 24 hours of the day for running the project successfully. Provision of Diesel generating set of the capacity of 330 KVA is thus a must. Water supply of 150 KL/day is the optimum requirement for this size of the unit. Flowers are of perishable nature require refrigerated transport to the Cargo center at the Air Port. Transportation of the staff to the unit and back is the other requirement for the Utilities head and forms a permanent part of the project costs.

Irrigation and Fertigation system has been designed for Sand Bed trenches, 12160 meter in length, 0.50 meter wide and 0.90 meter deep with Drippers to be provided in a length of 24320 Meters. Irrigation water will be subjected to filtration for sand and silt particles and Anti-Calcium treatments. EC (Electrical Conductivity), pH, Humidity, Temperature and Wind sensors will control aurtomatic working of computers for fertigation, Irrigation, Climate. Climate cooling is to be controlled by Fan and Pad system during summers and raising of winter temperatures with Hot air generators.

MISCELLANEOUS OR UNFORESEEN EXPENSES: This head covers all other expenses not referred so far and falling within the ambit of Project costs are included in this head.

Estimated Share of Costs Head Wise

- Land cost and Development 5%
- Plant and Machinery 50%
- Civil Structures 5%
- Duties, Fee's, Transport,
- Materials, Fixed Assets,

- Contingencies and Know-How. 20%
- Reefer Van & Other staff Transport 5%
- Miscellaneous 10%
- Working Capital 5%

	Rupees in Lakhs
TOTAL PROJECT COST :	540
COST PER SQ. METER :	0.27
COST PER 1000 SQ. METER	27.00

	Expenses Sub-heads	***Cost/ Unit***	***Percentage of Project Cost***
1	Green House complete with structure, Polyethylene film cover, Electric curtains, Insect Proof Net	99	18-20
2	Irrigation system complete with drippers, sprinklers, Electric valves and accessories.	5	1-1.2
3	Water Filteration system complete with Back Flash, Sand separation, Anti-calcium treatment and accessories	6	1-1.5
4	Computerized controls for irrigation, fertigation, EC and pH sensors, soil Humidity sensors, communication elements and software.	17	3-3.3
5	Climate & Environment control with Electric Fans, Cooling Pads, Air Mixers, Sprayers, Insect Killers, Sulphur evaporators, sensors for temperature, humidity, wind, radiation, Electric boards and cabinets, software.	67	12-13
6	Drainage Pipes	7	1-1.5
7	Cultural Tools	8	1-1.5
8	Rose Plants	89	15-17
9	Contingencies	22	8-10
10	Engineering & Planning	15	2-3
11	Assembly & Inspection	3	0.5-1
12	Plant Freight / Transport to Indian	10	1.5-2

	Port		
13	Know-How Fee	26	4-5
14	Land Cost	25	4-5
15	Land development for Roads, Green House erection, Levelling and Provision of Tube well etc.	10	1-1.5
16	Civil construction	36	4-7
17	Reefer Van and other Transport	30	5-6
18	Customs Fee's, Power connection, Diesel Genset, Fixed Assets like Cold store, Cool room	20	3-3.5
19	Pre-operative Expenses	15	2-2.5
20	Miscellaneous expenses		8-10
21	Working Capital	30	5-6

Detailed Break-up of Plant and Machinery Costs

Green house units will comprise of erection on Hot Galvanised Iron covered with Infra Red / Ultra Violet Polyethylene film, 150 micron thick, with a Total requirement of 41552 sq. mtrs. Each Green House unit of 5120 sq. mtr is to be fitted with 4 units of electric curtains complete with motor and components, thereby deploying 16 units in all. For the control of Virus vectors – Insect proof Net of 50 mesh 1164 sq. mtr. to be provided on the sides of crop in each unit. Total Net to be put in 4 Green House units works out to 4656 sq. mtrs.

	ESTIMATED COST	
	In USD	INR in Millions
Green House :		
1. 2.5 mm Hot Galvanized structure, 160 x 32 x 4 meter each for 4 units.	236940	10.6
2. Polyethylene Film 150 micron thick, 10388 sq. mtr., each kg weight of film covering 7.09 sq, mtr for 4 unit each of 5120 sq. mtr.	34627	1.56
3. Electric curtains 4 No's. per unit of 5120	11043	0.50

sq. mtr with Motor and accessories – Total 16 No. curtains.

4. Insect Proof Net 50 Mesh, for control of Virus vectors. 1164 sq. mtr for one unit of 5120 sq. mtr with a Total requirement of 4656 sq. mtr. — 11282 — 0.50

Irrigation — 23163 — 1.40

1. Total Bed Length in each Unit 3040 R. meters
2. Distance between each Plant row 20 cm
3. Distance from Plant to Plant 20 cm
4. Provision of 2 Dripper lines in each unit with Total line length for 3040 mtr. 2 x 3040 R. meters
5. Gross Dripper Length for 4 units
6. 2X4 x 3040 R. meters
   DRIPPER SPECIFICATIONS : 20 mm, with 1.7 liter per hour discharge, 20 cm distance between each dripper and 2 lines per row.
7. 70 liter per hour discharge Sprinklers, 815 pieces.
8. 2 inch Hydraulic Valves – 38 No's
9. Accessories for Irrigation system

Filteration and Cooling System — 14901 — 0.67

1. Filteration and Anti-Calcium treatment unit for Pad and Fan system, 4 units with control Filters and 1 inch Anticalcium.
2. Automatic Back Flassh system for irrigation 35 cu. mtr/ hour, 2 inch cleaning - One Unit
3. Sand Separator filters 3 Units
4.. Anti-calcium treatment 1.5 inch Two Units.

Computer Control — 50866 — 2.28

- Computerized Control Head – 2 systems
- Fertigation Pumps for Macro and Micro nutrients, fertilizer pumps – 3 for each computerized control head – 6 Units
- Communication Elements – 3 Unit
- Lightning Protection – 5 Units

- EC and pH sensors 1 No. per Head – 2 Units
- Soil humidity sensors 1 set for 1000 sq. mtr of 2 sensors of 30, 60 cm depth + Extractor. 20 Sets.
- Software connection to Central computer for Fertigation control.

Climate Control	298517	13.43

- Computerized control Head – 2 Units
- Cooling Pad system complete with – 640 Mtr. each for Pads and Distribution Panel and 640 Mtr. water distribution and collection Pipes.
- Electric Fans SS 1.5 HP, 48 inch – 25 No's per unit of 5120 sq. Mtr., 101 No's.
- Air Mixers 0.75 HP 24 inch 10 per unit for 4 units of 5120 sq. Mtr. – 41 No.
- Hot Air generators for Heating system - 200,000 kcal/hour complete with Chimney, Diesel Burner & Thermostat – 3 Sets per unit of 5120 sq. Mtr. – Total 12 units.
- Mechanical sprayers for Pesticides and cold Fog – 6 pieces.
- Sulphur Evaporators 0.1 kw / hour – 200 No's.
- Sensors for Temperature, Humidity and Wind, Wind speed and direction One Unit, Breathing cell for temperature and Humidity one for each Green House – 4 units.
- Electric Boards and Cabinets one each for each Green House – 4 units.
- Software connection to Central PC for Climate control – 1 unit.

Optional Systems : For maximum efficiency of cooling & Heating, Electrically operated	132025	6.00
Aluminium Net : Manual operated	39500	1.78
Cultural Tools : for pruning, cutting, hoeing and weeding etc.	6330	0.30
Rose Plants : 130,000 No's. @ 6.25	260,000	11.70

	Plants per sq. Mtr.		
	Contingencies : 7.5%		
	Engineering and Planning	20,000	0.90
	Freight to Bombay Port	68750	31.00
	Assembly Inspection	7500	0.34
	Know-How Fee's	80,000	3.60
Local Machinery and Equipment			
1.	Diesel Genset 2 KVA – 1 No.	26000	1.17
2.	Electric Boards, Panels, Wiring Poles	25000	1.15
3.	Fertilizer Tanks, 5000 liter capacity – 2 No. and 1000 liter capacity – 1 No. @ Rs 2.80 per liter.	2000	0.09
4.	Cold Store, Pre-cool Chamber Machinery for cooling to minimum of 2 °C.	15000	0.68
5.	Reefer Vans – 2 No's. TATA – 407	40000	1.8
6.	Sand Bedding - 12160 X 0.5 X 0.90 or say 5630 cu. Mtr.	25000	1.15
Civil Structures			
1.	Buildings 35 x 20 sq. Mtr area in RCC pillars with single brick (9 x 4½ x 4) wall, plastered, accommodating packing shed, cold store, Pre-cool Chamber, store, Office, Generator Room, Gutters for 4 Green House Units.	85800	3.90
2.	Furniture for Office, Packing shed, Racks for Pre-cool and Cold store Chamber.		0.2
3.	Cost of Land and Land development		3.5

FLOWER PLANT SELECTION

Selection of the flower is done on the market requirement whether for the cut, decorative or dried purpose. Ideal flower selection should have the characteristics of high market value, high production per unit of area, extended production periods and marketing season, long productive life span, vase life, resistance to drought, pests and diseases, possessive of aesthetic look with foliage and stems, ability to be sold as fresh. All the flowers have not all these qualities contained in them but the producer must keep these criteria for the selection of the flower for floriculture of commerce in mind. The two classifications thus arise are the Annuals and Perennials.

CLASSIFICATION OF FLOWERS

Floriculture plant classification for cultural aspect of the plant can be done on Thermo-sensitivity of seasons and also their nature of growth. Classification of flowers has thus been necessitated on the type of flower. Flowering plants can thus be divided into Summer, winter flowering and then the Annuals and Perennials.

Perennials : are plants that live year after year. Trees and shrubs are perennial. Garden flowers are herbaceous perennials, which means that the tops of the plant - leaves, stems and flowers die back to the ground each winter but the roots persist through the winters and new plants come up when the temperatures are high enough. Plants surviving such vagaries are termed hardy. Some of the perennials are not winter hardy but are classed with annuals, as these require setting in the flower beds each year. Examples are Begonias and Snap dragons. Perennials has a limited blooming period of few weeks only. Advantage of perennials is that these need not to be set every season like annuals.

Perennials are selected with a purpose in mind whether for edging or accent for evergreens or mass colour presentation or rock gardens in the home or for commercial markets. By choosing the purpose plant selection is made in the home gardens. Cultural management requirements like setting out the plants out doors, watering, mulching, weeding, fertilizing, disbudding, staking remain the same.

Annuals : Annuals complete their life cycle within one year and usually bloom through out the growing season and produce seeds. Positive features of the annuals are that sturdy, easy to grow, produce instant colours. However the greatest disadvantage is that the old flowers on the plant tend to die and require constant removal if kept for colour, other wise the plant will go to seeding and die. Some of the annual flowers has the ability to shed the seeds which come up naturally the next year. Alyssum, Petunia, Cosmos and Impatiens are some amongst this category.

Annuals are seeded every year. Annuals must be selected and placed in the garden as per the plant character of height, form, colour. Annuals are deployed only as fillers till permanent plants take their place. Starting culture of annuals for the garden, seeds are first raised in the nursery beds and kept well protected from heat and cold during summers and winters. After this starts the process of plant setting in the garden watering, fertilizing etc.

Bulbs/Corms/Tubers/Rhizomes : These are the modified developed stems or buds of the plant where in the food is stored in the form of leaf sheaths, fleshy under ground stems, or in root like modified forms from the underside. In common usage a Bulb may mean a Tuber, Corm or a Rhizome. Major flowering plants of the Bulbous group are Achimenes, Crinum, Gloriosa, Haemanthus, Hemerocallis, Hedychium, Hippeastrum, Hymenocallis, Zephyranthes, Dahlia, Freesia, Gladioulus, Narcissus, Canna, Ranunculus, Crocus, Cyclamen, Agapanthus, Alstroemeria. Popular bulbous foliage plants in the group are Anthurium, Caladium, Maranta, Monstera, Alocasia, Colocasia. In contrast all have common factors of food storage, rapid growth, same life cycle i.e. growth, flowering and dying away after maturity as that of the perennials.

BULBS : Shortened underground stems Bearing overlapping membrane like leaf bases functioning as storage of food for the plant i.e. Onion, Leek, Shallots, Chives, Garlic.

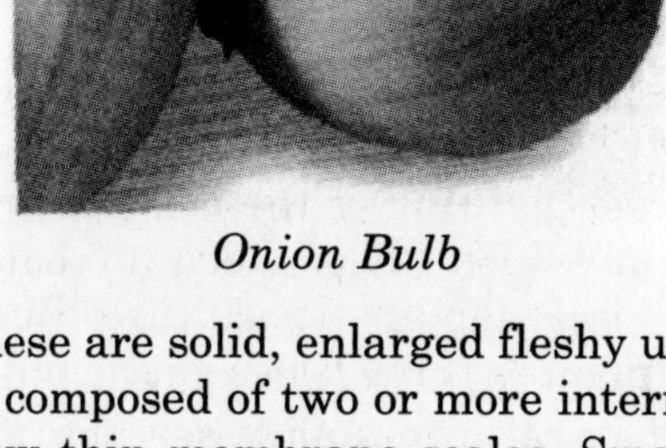

Onion Bulb

CORMS : These are solid, enlarged fleshy underground stems rounded in shape composed of two or more internodes and covered externally by a few thin membrane scales. Small corms attached to the main corm are called cormlets.

TUBERS : Fleshy underground stem with dormant eyes, specialized for food storage and capable of giving rise to a new plant such as Potato, Jerusalem Artichoke.

Artichoke Tubers

RHIZOMES : A stem root like in appearance lying on or creeps under the ground, which roots and sends shoots, often scale like in appearance, such as Ginger, Turmeric, Arrow root.

Turmeric Rhizome

Ginger Rhizome

Forcing : Practice of encouraging plants to flower or fruit early. In the case of bulbs this is the practice to remove the dormancy, affected by subjecting the bulbs to cold treatment. Importance of forcing lies in development of good root system before quick growth.

Planting : In the planting of the bulbs a prior knowledge of the correct depth at which these need to be placed must be acknowledged as a vital step for floriculture. Bulbs are generally recommended to be planted three times as the bulbs are wide.

Storage : Bulbs are harvested when the plants have died down and the foliage dried. Immediately after harvest the bulbs are trimmed, prepared for spread under the shade to dry down.

After a week or so the clean bulbs or corms are placed in cool well ventilated place. Inspect occasionally the stored produce for disease or mold appearance.

Cut Flower : Speciality cut flowers of Roses, Gladiolus, Lily, Narcissus, Dahlia etc. in the markets of Delhi and other metropolis of the country is very recent phenomenon. Farmers began to diversify into high – value speciality flower crops with the growth in demand for these. Retail florists also began asking for the culture of these new flower types and resulted in the coming up of very many Farms closer to the Delhi capital. This resulted in the exchange of information between the farmers and the scientist and thus new seed stocks imported from Europe for these saw a dramatic turn to cut flower business. Economics of cut flower production even in the smallest gardens is a profitable venture today, because of the increased and yearly growing demand for flowers. First time producers are advised to start small and learn the techniques of management in the first year of establishment and extend the purpose of marketing to the second year. This is for the rural areas very near or adjacent to the urban market. Jasmin culture in the South India, Marigold or Tagetes in the North, Roses and Jasmin in the West Bengal are some of the examples of home flower culture in India. Potential markets exist in the metropolis of these regions. In Delhi the Government has now come up with an auction and specialized marketing center for flowers in the South of Delhi for the benefit of the farmers and giving an encouragement to the traders. Flower production must be based on the type, variety and the requirement of the market, on which depends the designing of the production system.

Marketing : Flowers are the highly perishable commodity and requires immediate auction or good post harvest handling technique to gain access in the far off markets of the country or for export. Marketing of cut flowers can be done either to the wholesaler, a retailer or directly to the consumer. In the wholesale markets, own standards are set for each type of flower which comes to the market and the producer must adhere to this, in order to gain maximum access and easy sale of the produce. Prior to 2000 flower markets in Delhi had the specialized flower sales i.e. Mariglod and Jasmin in the Chandni Chowk Markets and the roses in Khan Market and Defence colony markets. With latest decisions of the state government of Delhi, whole sale market has been created in the Mehrauli area of Delhi.

Site Selection: Site selection where floricultural operations are to be started must conform to certain pre-defined requisites of a profitable venture. Sun light, soil, water for irrigation and winds in the area form the prime factors for site selection. Flower plant selection is based on these. Sunlight is whether full (8 hours) or partial (5-7 hours) or partial shade or shady. Structure and texture of the soil whether sandy/loam/clay, fertile and well drained. Quality and availability of water. Openness of the area or presence of trees and shrubs further presents the picture for the winds in the area. These factors play an important role in successful floricultural operations.

PRACTICAL TIPS FOR SUSTAINABLE FLORICULTURE

1. Start small.
2. Floriculture is highly labour intensive and may present challenges during the first 2-3 years.
3. Floriculture marketing takes much of the time than producing.
4. Plan what to grow, where to grow and how to market? Planning involves the study of the market, its demand for the type of flowers and the product whether it is for flowers in bunches or for Bouquets and what are the colours most highly in demand. Bunch sales in the market can be for Zinnias, Sunflowers, Dahlias, Chrysanthemums, Gladiolus, Lilies etc. Bouquets demand growing of stems with lots of small flowers and with varied colour combinations. Clientele also matters, when women are the purchasers then light sober colours of pinks, blues, purples are desired. In the case of men the colour preference changes to hot colours like Red, Orange and Yellows. In this planning, while deciding what to grow, consider what makes a good cut flower :

    - Long vase life
    - Good colour and colour retention
    - Long harvest time.
    - Low cost production.
    - Large number of stem production.
    - Good stem length.

- Flower presentation on the stem where the bloom is located.
- Will the selected flower is good for the region.
- What is the quality of resistance for pests and diseases?

5. Production methodology to be defined whether open field or protective culture. This all depends on the labour cost and ultimate production costs in the area. Important Flower species in the perennial group are Yarrow, Globe Thistle, Lily (*Lilium* spp.) Rudbeckia, Veronica. In the Annuals group these are Ageratum, Snapdragon, Celosia aristata (Crested Cocks Comb), Celosia spicata (Plumed Cocks Comb), Larkspur, Cosmos, Sweet William, Lisianthus, Gomphrena, Statice, Salvia, Marigold, Zinnia etc. Though these are the most popular species but the list is by no means complete. In this another point of mention is to grow those flowers in the area which no other farmer is growing and the introduction is not only new and fascinating but poses no competition for its marketing.
6. Schedule the production in such a way that most of the big holidays and occasions do get the cut flowers on demand when these are most needed. This can be achieved by extending the growing season which can be done either by early culture or making use of Tunnels (Igloo's), growing fillers, using cultivars if different colours and maturities. Another notable feature is that while annuals take only a season to grow, perennials take 2-3 years to produce saleable stems.
7. Integrated Pest Management demands continuous monitoring of pests and developing thresholds, using spot treatments that are least harmful to beneficial insects, the environment and well accepted in the market. Alongside build up populations of beneficial organisms for pest control and attacks on the crop. Make use of sticky cards or traps for pests.

Weed Control : For optimization of productivity, weed control is a long term process to keep these under control so that these do not compete in any way with the crops. In the sustainable floriculture, a single method is not successful however integration of all the methods at disposal can restrict the weeds and keep these in control. Various practices done by the farmer are crop rotations, interculture operations, mulching, cover crops. Latest other methods are the Solarization of the soil and use of Bio-herbicides. Mulches can be used in flower production including cover crops, plastic, paper, organic materials like grass or straw, bark and hay. Mulches are not only advantageous for weed control but conserve moisture, restrict soil erosion.

Latest introduction in the weed control is herbal or Bio-herbicide. Corn gluten meal (CGM) is a by-product of Corn processing, which has proven its worth as a bioherbicide. CGM is a pre-emrgence herbicide applied in early spring. It works best when applied to the top 1/4 inch of the soil. CGM contains 10% nitrogen and acts as a slow release fertilizer for the crop. CGM is a patented product and is currently being sold in Europe as a herbicide. Latest researches has shown that Corn gluten hydrolysate (CGH) which is made from Corn gluten meal is more effective than CGM and could be applied at half the cost of CGM.

Organic Manures : Bio-degradable materials of the animal and plant origin, agro-wastes of the farms and crop culture are all covered as organic manures. Range of organic manures available in the market and what a farmer can produce in his field is too many good manures these days. Compost from agro-wastes of the farm, animal manures (Cow, Sheep, Buffalo), Poultry manures, de-oiled cakes, vermicompost (decomposed organic wastes by Earth-worms), animal meals (Bloodmeal, Bonemeal, Agromeal), decomposed wood shavings and saw dust, tree bark. Farmers can use these in place of inorganic fertilizers. These are not only cheap and highly nutritive supplements but environment and soil friendly. Organic manures are complete slow release, contain micro-nutrients needed by the crops. Organic manures contain all the macro and micro food materials required for plant growth. These enrich the soil by supplementing the plant food requirements of the crop and also build a healthy microbial activity in the soil for plant growth.

Water : Water is the single factor in floriculture and cut flower trade and fresh cut stem right from harvest till delivery to the consumer. Water used is from different sources i.e. rivers, lakes, reservoirs, tube wells, municipality. Apart from water source, water quality is different every where. Thus water source and quality can have huge difference in pH and chemicals and also purity, contamination. Water contaminated with organic matter and micro-organisms interfere with chemical additives and blocking of water movement in the stem. Water has both positive, negative effects on the vase life. In water another factor is the presence of air which has the power to affect the balance of shelf life. Water with less air results in higher uptake. Thus water is a key factor in flower culture performance during production from harvest till delivery and increasing vase life. Cut flower stems respond best in water with a low pH 4.0-5.0. In this pH range not only water uptake improves, but effectiveness of biocides also gets enhanced. tap water has a neutral pH or above 6.0-7.0 and needs acidification to bring pH to the desired levels. However precaution is observed for Gerbera and Chrysanthemums, which suffer from discolouration due to low pH.

Toxic Elements : Toxic elements like chlorine, fluorine, boron and iron are some of the toxic elements in water for cut flower stems. This toxicity can also be there with Zinc buckets. For this always use plastic buckets. Damage to Chrysanthemums with fluoride levels as low as 5 ppm can occur. Freesias, Gladiolus, Gerbera, Tuberose are damaged due to fluoride which can be possible even at the lowest levels of 1 ppm fluoride.

Why use pH Water Treatment : To guarantee customer satisfaction for vase life, this relies only on water treatment of cut stems after Post Harvest. Treatment ensures longer vase life when treated correctly. Consumer buys more when he is sure of his money's worth. Many growers has their own product formulation for pH treatments, but it is highly feasible and recommended to use only proprietory products from the market which are treatment specific but also crop specific. For this proprietory products both for hydration and holding of cut flower have become much more popular. Some of the products in this are "FLORA LIFE's" hydration and holding solutions in the ready to use and concentrated form.

Common Insect and Pests : Pest problems may vary from year to year but some of the common crop pests are Grasshoppers, Spider mites, Thrips and Aphids. Biopesticides which are not only cheap and easily available and can also be prepared in the home like Tobacco decoction, Neem oil sprays.

Common Diseases : Knowledge of diseases and their control is another part of IPM. Most common diseases of the cut flowers are Botrytis, Powdery mildew, Root rots.

- **Botrytis :** This disease occurs in cool, wet and humid weather conditions. This may affect any plant or plant part. Diseases mostly attack ageing or dying flowers and seedlings. Alternative controls for this are use of Baking Soda preparations or Bordeaux mixture spray. Baking Soda preparation acts as a fungicide. Take 4 TS (18-20 gm) of Baking Soda and 5 TS (25 ml) Horticultural oil and mix this with 4 liter water. Cultural controls are use of resistant varieties, hand picking of pests, avoiding over irrigation, controlling poor drainage, crop shading and good practical knowledge and application of companion cropping.
- **Powdrey Mildew :** Conditions which favour the development of this disease are warm, sunny days and cool nights with high humidity. Powdery mildew is not a problem in temperatures above 18°C. Spores of Powdery mildew are carried by wind. Flower plants most affected by the disease are Achillea, Anemone, Aster, Brachycome, Companula, Chrysanthemum, Cosmos, Dahlia, Delphinum, Dianthus, Gazania, Lathyrus, Phlox, Solidago and Zinnia. Alternative controls for this are Neem oil, Sulfur and use of parasitic fungus (*Ampelomyces quisqualis*). In the cultural side removal of infected plant parts, host weeds and rotation, keeping soil fairly dry etc.
- **Root Rots :** Root rots are encouraged by wet soil. Secondary controls recommended are the use of parasitic fungi *Gliocladium virens*, parasitic bacteria *Streptomyces lydicus* for control of Pythium and Rhizoctonia root rots, fungus *Trichoderma harzianum* for all root pathogens. Cultural management recommended technologies are steam heating or Solarization of the soil, use of compost and avoiding over irrigation.

Post Harvest Management : This is the most important process in the life of the crop and starts immediately after harvest of the flowers. Flower harvest is recommended to be done in the coolest temperatures during the early hours of the day or in the evening. During the noon hours when temperatures are very high harvesting be not done. Cut flower stem ends are immediately submerged in water buckets. The technology enshrined in the process envisages delivery of the freshest cut flower stems to the consumer in the shortest possible period without any appreciable deterioration. The process involves quality maintenance, increasing the vase life, temperature, moisture loss control. Optimum temperature flower specie specific is maintained with high humidity and supplementation with floral additives for stem nourishment and bacterial control, in the cold store chambers and transporting reefer vans to the auction house. Apart from this ethylene sensitivity of the flower is observed by avoiding stocking with ethylene producing materials or plants.

Table: Showing Problems of Cut flowers During Post Harvest Storage and Marketing

Flower	Major Problem			
	Ethylene Disorder	Water Stress	Leaf Discolouring	Remedial Measures
Agapanthus	Yes	No	No	STS
Alstroemeria	Yes	Yes	Yes	STS
Carnation	Yes	No	No	STS
Chrysanthemum	No	Yes	Yes	Clean solution
Daffodil / Narcissus	No	Yes	No	Clean solution
Delphinium / Larkspur	Yes	Yes	Yes	STS, Clean solution
Freesia	Yes	No	No	STS
Gerbera	No	Yes	No	
Gladiolus	Little	Yes	No	STS
Gypsophylla	Yes	Yes	No	STS
Iris	No	No	No	
Liatris	No	Yes	Yes	
Lily	Yes	Yes	Yes	STS
Daisy	No	Yes	Yes	
Protea	No	No	Yes	
Roses	Yes	Yes	No	STS
Snapdragon	Yes	Yes	No	STS

Statice	No	No	Yes	
Sweet Pea	Yes	No	No	STS
Sweet William	Little	No	Yes	STS
Tulip	No	Yes	No	
Wax Flower	Yes	Yes	No	STS
Anthurium	No	Little	No	
Bird of Paradise	No	Little	No	

Source : Flora life, Inc – Horticulture Technologies, Holland

STS: Proprietory products Silflor/50, Silflor/RTU for usage against Ethylene sensitivity.

CHAPTER - 6

INPUTS AND PROCESS FOR FLORICULTURE

Flower culture has two distinct facets in the cultural process. These are Pre-Harvest and Post-Harvest processes, where inputs are very different from each other. In the pre-harvest the type of culture, fertilizers, irrigation, IPM (integrated pest management), weed control, planting and spacing, time of sowing are the aspects. In the Post-harvest it includes the product handling at harvest and after harvest and preparation for marketing, storage and transportation to the market under a complete process of cold chain.

Green House : Green house covered with UV protected Polythene film with an individual chamber of 5120 sq. mtrs in area, fully equipped with drip irrigation, water filtration, fertigation, soil moisture sensors, Pad cooling system for temperature control, pest nets is thus the best technology.

Civil Structures : Civil structures will include const-ruction of an integrated system for cooling of the harvested flowers in the form of Pre-cooling chamber and a cold store, Packing and Grading hall.

Reefer Van : For the transport of the processed, pre-cooled, packed product and the produce for Air lifting from the International Airport the product needs to be transported to the Cargo center without breaking the cold chain.

Process : Floriculture process starts immediately with identification of the type of culture. The process involved in this is monoculture of Roses only. The selection is based purely on the recommendations of the Government and the suitability of the areas in particular selected zones of culture.

Selection of Varieties : Selection of varieties is the process based on the market demands for a variety, preference of the Importers and price value it commands in the market

CULTURAL FACTORS

Growing Media : The most vital input in the covered culture is the growing media. Due to shallow depth and volume of the container or artificial trench, growing media should be amended to provide physical, chemical properties necessary for plant growth. Field soil is not good to provide aeration, drainage, water holding capacity. Soilless growing medias which have been necessitated by the innovations of covered culture have been developed in the western countries from Peat, Reed and Sedge, Sphagnum Moss, Wood residues (Leaf mold, saw dust, Tree bark), plant wastes like bagasse, Rice hulls, Minerals - Perlite, Vermiculite (expanded quartz), clacined clay, Urea formaldehyde etc. In the preparation of growing media mixtures, the basic materials like Peat, Sphagnum Moss or Reed and Sedge should form a major component in the composition and Mineral materials should form only 25-33% of the constituted mixture. Vermicomposts, Composts also are the basic materials.

Growing Media Qualities

- Porous, well drained, moisture retentive.
- Low in soluble salts, with high exchange capacity to retain and supply the essential plant growth nutrients.
- Free from harmfull soil pests and pathogens, nematodes and weed seeds.
- Following pasteurization or solarization should be biologically and chemically stable, free from organic matter that releases Ammonia when subjected to heat or chemical treatment.
- Growing media should be standardized, uniform to act for a fertilizer program as a standard material.
- pH of 5.0 is optimum for soilless growing media.

Relationship between the growing media, air and water is least understood with the masses and in the commercial production of Green house and nursery crops. As a result there can be significant amount of plant loss either directly or indirectly due to basic understanding of the factors like air, water and the growing media together. Pore space in the growing media is the major criteria in determining the size and distribution of pores. Soilless growing media contains 60-80% pore space. Portion of this space is occupied with air. Plants require Oxygen, so adequate aeration in the media is necessary. Pore size determines the rate of drainage and gas exchange, which has a influence on the water holding capacity of the media. Availability of this water is determined, how tightly this is held by the particles.

Light : Shelf life of most cut flowers improves with rising light intensity as this increases the efficiency of photosynthesis and the carbohydrate content of the flower. Low light intensity results in the elongation of the stem and petiole. Low light intensity also leads to problem of bent neck in Roses and Gerbera and stem bending in Carnations, number of flowers on a stem in spray varieties with a weak stem. Apart from this low light intensity levels produce paler, poor quality blooms. High levels of light also have the adverse effects like bleaching, red coloration of tissue, leaf spots, leaf browning, petal droop etc.

Temperature : High temperatures are associated with rapid water loss and increased stress on plant reducing the quality and shelf life of the bloom. High temperatures reduce the bud size and the number of petals, which result in shorter stems. It is thus imperative to maintain perfect Cold chain conditions for a particular cut flower right from the Pack House till the destination.

Humidity : Humidity is another cultural factor which plays very important role and is very frequently under-estimated. Humidity fluctuates very much during the day and in rainy seasons and during sunny days when levels are very different. Humidity very much influences the rate of transpiration and uptake of nutrients and the development of injurious diseases and pests. High humidity encourages grey mould and low humidity encourages powdery mildew. High humid conditions improve leaf colour and the condition of the stem and bloom. Long periods of high humidity reduce transpiration throughout a crop and induces mineral deficiencies like Calcium causing problems like leaf tip burn, pre-mature petal drop.

Fertigation : It is the process of applying fertilizers along with irrigation. Excessive use of Nitrogen leads to the elongation of vegetative growth and the colour of leaves but shortens Vase life and encourages disease infection. High Potassium doses encourage hardening of stems and petiole, prolong shelf life and discourage diseases. Over watering causes minor plant nutrient deficiencies of Iron and manganese and may render flowers un-saleable. Water stress both under and over watering are the cause for reduced shelf life and poor blooms. Over watering reduces soil Oxygen levels and ultimately restrict root activity and causing loss of vase life and flower quality.

Chemicals and Phytosanitation: Usage of pesticides and chemicals that conform to the conventions and obligations of International trade with specific requirements of a country is a must. Pesticidal residue levels is becoming a major global concern. Thus pesticide residues on the leaves, petioles, stems greatly reduce the over all quality of the bloom. Diseases, rotten and decaying materials, smoking etc. leads to the production and accumulation of Ethylene in the pack house and cold store area and is harmful for the quality and shelf life. Bacterial entry into the cut wounds and bruises affects the ability of the flowers to take up water and reduces the vase life. Tradesrs in Netherlands are very particular in the Post Harvest process where sanitization is given a preferential place for the quality, shelf life maintenance. All materials, machines are kept scrupulously clean and sanitized and disinfected on daily basis.

POST- HARVEST MANAGEMENT

Flowers are very sensitive to any type of stress right from the time of harvesting or flower handling till it reaches the final destination. All handling procedures thus have a pronounced effect on the flower quality. Prior to the decade of nineties the flowers were judged on the basis of colour, flower development, appearance, stem length and quality of the leaves. Now the quality of the flower lies in its Vase life potential. Emphasis of to day has thus shifted to the ability of the cut flowers to last longer when in a Vase. Though variety, root stock, basic cultural practices like fertilization, irrigation, pest and disease control affect the final bloom but Post Harvest Operations have become the most important factors in the life of the flower and thus the Vase life. It is thus very important to understand the basic underlying technology right from the harvest to pre-cooling, grading, packing, hydration, preservation, storage, transport/shipment from the farm up to the destination center.

POST-HARVEST AND HANDLING

Harvesting : Exact stage at which a flower is harvested is very critical and well defined for a particular flower. Roses are to be harvested when the first petal is reflexing. In the case of Chrysanthemums, Gerbera, Aster which are the members of Compositae need to be cut when fully open but before the stamen produces pollen. Statice is harvested when all the flowers in the floret are just open. Growers should not get tempted from the trade for better prices at the cost of flower quality, but make sure that flowers are harvested at the proper maturity level. While harvesting damage to flowers, stems be avoided from causing bruises and crushing etc. and operation is started only in the cool hours of the morning or late in the evening. Careless cutting or tearing leads to un-necessary bleeding and quality loss. Various stages at which a flower is to be harvested is as under :—

Flower	Stage of Development and Harvest
Alstroemeria	4-5 florets open
Chrysanthemum	Fully open
Carnation (Standard)	Half open
Carnation (Spray)	2 fully open flowers
Dendrobium	Fully open
Gerbera	Outer row of stamens just forming pollen
Gladiolus	1-5 buds showing colour
Rose	Calyx reflexed, first petal reflexed

Handling: Harvested blooms are needed to be immediately placed in cool water in a bucket or collecting bin. Cut blooms are shifted to the cold store for pre-cooling. Immediately after pre-cooling the process of grading, bunching, packing and preparation for shipping and transporting starts. Period from harvesting to cold storage should be as short as possible. As cut flowers are a living organ and continue to grow. The process of growth is arrested under controlled conditions of temperature and humidity levels.

Cold Storage : Very specific temperature conditions are maintained for different flowers. It is thus imperative to take the temperature of the crop and not that of the air, which is most significant. Even as low a temperature difference of ± 2°C is significant to accelerate the maturing process. Effect of Ethylene in the cold store is another vital factor having effect on the vase life and flower quality. As rotting, decaying, ripened fruits of Melon, Apple, Papaya, smoking can create conditions of Ethylene production and needs to be curbed and avoided. Cold stores should be thus ventilated for the exhaustion of Ethylene at least once in a day.

Pack House Process : Pack House design and layout are the critical inputs of a Pack house for cut flower processing. For one hectare of Flowers minimum area of 300 sq. mtrs is recommended. Pack house further should be temperature controlled, with good many grading tables covered with clean and sanitized material. All grading tables are made according to the sizing requirement of the flowers, so that minimum time is taken in cutting and bunching of flowers. Tables should also have the arrangement of fixing buckets which remain handy for the grader to pick up flowers.

Grading : Cut flowers are graded according to the desired specifications for the specie and requirements of the Importing country. Cut flowers are classified into grades such as "EXTRA, 1st, 2nd, 3rd". These grades are generally given codes such as Gold, Silver, Bronze. Criteria for each grade and bunch is as under :—

- ***Stem Length*** : Longer the stem better the quality.
- ***Stem strength*** : Stronger the stem higher the quality.
- ***Number of Flowers*** : Standard varieties should have one flower per stem, spray should have a specified number of flowers and buds. Cymbidium Orchids are priced for only number of flowers per stem.
- ***Development of flower*** : Here the bloom is seen whether it is in tight bud or half open or mature condition.
- ***Appearance*** : Flowers should be well formed and free from any defect such as calyx splitting or bull heads or bent necks. Individual flower stems should have a good overall appearance and there should be balance between flower or the flowers and the stem. Take care that there is no dirt or deposit which detracts the stem.

- ***Pests and Diseases*** : All the flowers should be free from any pest or disease.

Packing : Packing is a combined process of three stages i.e. Bunching, wrapping and packing into box's. Graded flowers are bunched together in desired numbers, as per graded stem length. Each Bunch thus must contain the stipulated number of stems 5 or 10 or 20. Bunch's are put together without any damage to the stems or blooms and the leave and are secured by a rubber band. Jigs are also used at very many pack houses to reduce the timing or the extra exercise of bunching. Polyethylene sleeves are being commonly used these days for wrapping. However Roses are seldom packed in sleeves as the dense Rose foliage gets heated up quickly after bunching. Roses are thus packed or head wrapped with soft corrugated paper 15-20 cm wide and fastened with an elastic band taking care that wrap protrudes a little above the Rose flower buds. Bunched flower packs are then packed in box's having varied dimensions and weights for each category of flowers. Maximum per box weight for Roses, Chrysanthemums, Carnation, Gladioulus is 17, 15, 13, 15 kg respectively. Large returnable box's are also being used in the trade carrying a weight of 50 kg and capable of holding 1000 – 2000 flower stems.

Cold Storage : Optimum storage temperature and duration of storage varies with each individual flower. However as a general rule flowers originating from temperate regions i.e. Roses and Carnations are best stored at a temperature just above the freezing point of plant tissue i.e. at 2°C Flowers originating from tropical regions i.e. Anthuriums and Dendrobiums are sensitive to chilling injury and are stored at 10-12°C keeping RH at 90%. These conditions are based on the recommendations for Netherlands growers and by US department of Agriculture.

Occasionally an export grower needs temporary storage for pre-cooling, grading, packing and transporting into a Reefer Van for shipment at the Cargo complex and requires the very recommended temperatures. Storage temperatures with RH of 90% for different flowers are as under :—

Flower	Storage Temp. °C.	Maximum Storage Time in Days	
		Dry Packed	Wet Packed
Roses	0 – 1	3 – 7	7 – 14
Carnation	0 – 3	10 – 14	7 – 21

Chrysanthemum	0 - 3	10 – 14	7 – 28
Lily	0 – 1	3 – 4	4 – 7
Gypsophylla	4	1-2	2 – 3
Gladiolus	3	2-3	6 – 8

(**Source** : ITC-1997 : Product and Market Development Cut Flowers)

Preservatives and Stem Treatment : De-hydration and vascular blockade is the common phenomenon and source of damage in the cut flowers. Preservatives are used to protect the tissue from the bacterial, fungal, algal damages. Diseases produce or induce toxic substances such as tannins, which can block the vascular tissues in the stems causing pre-mature wilting and shortening of vase life. Emphasis these days is to inhibit such contaminations and prolong the vase life. Netherlands market has made the application of preservatives mandatory. Proprietory products are now in the market as bactericides, Ethylene inhibitors, sugars and plant hormones as growth regulators.

- Bactericides are made from Citric acid, which reduces the pH and encourages faster water uptake and is diluted at a rate of 100 ppm for soft water and 350 ppm for hard water. Hypochlorite diluted at 1000 ppm is an excellent and cheap bactericide for standing solutions and for general sterilizing of tables, water buckets.
- Cut flowers need carbohydrates to continue with vital functions such as respiration. Cut flowers with considerable foliage are less benefited. House hold sugars at the rate of 1.5% dissolved in water is ideal for the purpose. Sugar provides a good breeding ground for micro-organisms as such sugar treatment must follow bactericide also.
- Silver is an inhibitor for the production of Ethylene. Silver nitrate which was previously being used has since been stopped because of its slow translocation has been replaced by Silver thiosulphate (STS) which is quick in its action. Two types of STS treatments are recommended. Short treatment by keeping blooms for 10-20 minutes in warm water to which 40 ml / liter of STS has been added. Overnight treatment is accomplished to blooms at 2°C in cold water with 10 ml / liter of STS.

- Specialised corporate bodies in the Post Harvest technology of Flowers have come up with their products for the Ethylene factor control and hydration of the cut flower stems in order to provide a ready food. Floralife products in the markets of Netherlands and also their marketing in other world countries in the business of floriculture has made a remarkable contribution. Silflor is a STS product and Hydraflor is a nutrient supplement product, much in demand for the floriculture industry.

Transportation and Shipping : For shipping and transportation of the box's the pre-requisite is the proper labeling of the box's. Labelling describes the product origin, type and variety of flower, grade, quantity of stems and weight and other valuable data required by the Shippers along with the identification of the producers and the Importers. The process is aimed for speedy dispatch and arrival of the consignments at the destination without any hassle and flaw. Proximity of the Airport Cargo complex matters very little for the Reefer container and the produce already handled within the cold chain norms. Prior to shipping of the consignment every finer details are put into order i.e. shipping date, flight details, airway bill number, Number of box's being freighted etc. Each shipment requires prior booking on the Airline through to the exact destination place without any transfer in between the route. Phytosanitary certificate, legalized certificate of origin, commercial Invoice are the necessary documents.

CHAPTER - 7

PLANT PROPAGATION AND NURSERY MANAGEMENT

INTRODUCTION

In gardening and fruit culture, plant multiplication and propagation are the integral parts of Horticulture management operations. Alongside is the nursery management for better upkeep and growth of the multiplied or propagated materials. Natures usual method is by means of seed. But in cultivated fruit/garden operational culture, plants do not come true from seed. This is true of deciduous plants which must be propagated asexually because they do not come true to seed. This makes it thus necessary to reproduce the desired fruit plants by methods such as cuttings, layering, grafting, budding, runners of a plant specie. Methods needed to be adopted for a particular specie are characteristic specific and will work for some fruits while other methods are needed to reproduce other fruits. In order to perpetuate the progeny of some useful plants with the same desirable traits, it is highly necessary to propagate vegetatively. After seeds, vegetative propagation is the other method of plant multiplication. In the present century another facet of Tissue culture regeneration has been added to it, where the requirement of vegetatively propagated plants is in millions. This is possible only through Tissue culture technique where one tissue leads to the multiplication of whole stock. Plant cuttings are made from Hard, semi-hard and soft wood plants. Grafting and Budding methods of propagation are used on tree's where a specific rootstock is desired or when cuttings do not root satisfactorily or do not develop a root system large to support a tree of the desired size. Vegetative propagation is undertaken by the following techniques :

Sexual Propagation: Propagation of the plants is either by seeds or vegetative means. Plants produced through seeds is sexual reproduction process. Apart from cereals, vegetables sexual propagation is also followed in the case of annuals (seasonal flowers) and most of the shade, ornamental tree's. Other plants which are commonly propagated in India through seeds are Palms and Cacti.

Vegetative Propagation : This is a method in which complete plant is produced from one vegetative bud or several such buds. This process involves cutting, layering, division, budding and grafting. Large number of ornamental plants are multiplied by one or the other method.

- **Cuttings :** Stem, Leaf, Root
- **Division :** Suckers, Runners, Stolons
- **Layering** Branch's
- **Budding**
- **Grafting**
- **Bulbs, Corms, Tubers, Rhizomes**
- **Tissue Culture**

Propagation is done only in periods when the plant is under active growth phase and the seasons are well defined for this. Spring and Mon-soon are the best seasons for propagation when the plant growth is active. Further to know much about the growth of the plant, indication of Sap flow in the stems and branch's of the plants is indicative of active growth period.

Mother Stock : In this process of cloning vegetatively, the characteristics of the mother plant is perpetuated in the progeny. The progeny thus is the true representative of the parent plant with the same characters in it. It is thus imperative to maintain proper mother stocks, which form the basis of the regenerating progeny of the future. These are the selected plants having disease resistance and freedom from pests and other qualities of the plant which are fruiting period, colour and size of the fruit. Dormant scion wood from the shoots that grew in the last season and vigorous in appearance is selected. Make sure the shoots so selected are insect and disease free and are true to type. Cut scion wood is then bundled and kept moist with sphagnum moss or wet towels or cloth and wrapped in a plastic bag. These cutting are then stored at 1-2°C and is ready for use as the scion wood material. For identification each bundle is to be marked and labeled properly.

Bud Wood Collection : Bud wood is collected from the current seasons growth. Clip of the leaves with a sharp knife, leaving about a ¼ inch of the petiole. At all stages bud is always to be kept moist both before and after clipping of the leaves. Where the supply is for 2 –3 days usage, always keep the buds moist and wrapped in wet paper and stored at temperature 1-2°C.

Cuttings: Cutting is a process by which a plant is produced by severing a vegetative portion of the plant and rooting it in a favourable medium under optimum conditions. Various such plant parts are Stems, Leaf, Roots, modified stems such as Corms, Rhizomes, Tubers, Bulbs and Runners. Raising plants by cuttings is the easiest and cheapest method available with the nursery man. Stem cuttings can be categorized into Soft wood, Semi-hard wood and Hard wood cuttings. Success of rooting depends on very many external and physiological factors. External factors are Season, Sunlight, Temperature, Photoperiod, Humidity, rooting media and aeration. Photoperiod relates to the accumulation of ample food materials in the cutting. Amongst the physiologic factors which have a bearing on successful rooting are presence of leaf or bud, nutrient level. Rooting media for cuttings is very vital because of not only it holds moisture but at the same has the capacity to drain excess water and permit good aeration for the formation of roots. Application or treatment of cuttings with a rooting hormone i.e. Napthalene acetic acid (NAA) or Indole Butyric acid (IBA) has provided best results. Rooting hormones are available in the market for soft, semihard and hard wood cuttings. Mist application in the propagation of root cuttings is a marvel achievement for success of propagation in ornamental horticulture. Even the most hard in rooting plants like Ashok (*Saraca indica*), Peltophorum, Thespesia, Magnolia were made easy by misting and application of hormones.

Stem : Stem cuttings are categorized as Terminal, Heel and Node cuttings. Stem cuttings are taken from half ripe shoots in the case of flowers / ornamentals plants. The length of the cuttings in this case is nearly 3-4 inches. The cuttings are then planted in the rooting media or already prepared nursery beds, either in the open or under covers in the shade houses or Green houses. Covered structures are further provided with added facilities of

Misting or Fogging for quick rooting of the material. This process reduces the transpiration of the cuttings and aids in quick formation of the cork/crown at the cut end for roots to come up. Whole process of rooting takes nearly 30-45 days at the most. To have a very high success rate of rooting in the cuttings these are treated with root forming hormones. These hormones are either in the powder or liquid form. Gardeners generally prefer powder form of the hormone. These hormones are different for fleshy, semi-ripe and hard wood cuttings. The best hormone is Indole Acetic Acid (IAA). Hard wood cuttings are generally taken for Forest tree species and presently Poplar and Paulonia fortunii are the two forest plants which are being multiplied very much and are very high in demand.

Leaf : Healthy well developed leaves of numerous plants provide a useful means of propagation. This is generally done in flowering and fleshy plants, shrubs and bush's. The leaf is removed from the parent plant and light incisions/cuts are given on the back of the leaf. The leaf is then fixed on to the moist compost, keeping it away from direct sunlight, but providing adequate warmth for growth along with moist atmosphere.

Roots : Pieces of roots 2 inches in length are taken from plants with fleshy roots and are planted in the rooting media for rooting. Examples of rooting in Carrots, Turnip, Radish is well known to the farmers. Another addition is the production of Chicoons from Witloof chicory which are the product of rooting from the roots. Plants such as Gypsophylla, Gaillardia, Salvia, Aralia, Lagerstroemia, African violets, Kalanchoe can be propagated by rootings.

Layering : Plants difficult to be raised by cuttings can be raised by Layering even without the help of hormones. In this twigs remain attached to the parent plant, so the supply of nutrients is continuously maintained for the rooting process. Layering is practiced both in the ground and air. Ground layering is the easiest method. Air Layering or Guttie is the most widely used method in ornamental plants. Shrubs, foliage plants and even horticultural and other tree's are raised by this method. While making layering rooting hormones can be used for quick process. Though Chrysanthemum is either propagated by suckers or cuttings but has been reported that this can also be propagated through layering process. Propagation by air layering not only shortens the growing period but produce better flowers. Amherstia, Carissa, Dracaena, Gardenia, Cestrum do well by layering. Litchi amongst horticultural plants does well by layering.

Bulbous Plants : Besides Bulbs this multiplication and propagating method includes in it the rhizomes, corms and tubers too. Plants like Tuberose and Hippeastrum produce both bulbs and bulblets and both are used for multiplication. Gladiolus produce Corms and cormels. Both bulblets or cormels do not produce flower in the first year of planting. This is related to very many physiological and environmental factors.

Grafting and Budding: are horticultural techniques used to join parts from two or more plants so that they appear to grow as single plant. In Grafting the upper part **(SCION)** of one plant grows on the root stock / system of another plant **(Root stock)**. Rooted cuttings of plants grow very poorly or does not exhibit good characteristics of the parent plant. As such the process of grafting is undertaken where cuttings root poorly. In the process of budding a Bud is taken from one plant and grown on another. Grafting and Budding are asexual or vegetative methods of propagation. The new plant which grows on the rootstock after Grafting or Budding has the exact plant traits of the plant it came from. Rootstock provides the characteristic of hardiness, drought tolerance or disease resistance. These techniques however are performed by well experienced nurserymen because it requires a very great skill. These are very time specific operations like the others detailed above. Grafting and budding is performed at very specific times when weather conditions and physiological stage of plant growth both are optimum. Budding and Grafting of the horticultural crops may increase the productivity, repair the damaged plants.

Why Grafting / Budding : Grafting and Budding of plants make these to achieve the desired goals through this technique of plant propagation. Desired goals can be :—

1. **Change Varieties :** Change of varieties can be done for optimizing cross pollination in the orchard. Certain cultivars of the trees are not capable of self pollination.
2. **Root Stock Advantage :** Advantage of rootstock is of utmost importance in acquiring the disease resistance or growth properties. French Crab Apple is resistant to Crown Gall and hairy root and Malling VIII and IX are

dwarfing stocks in Apples, these are thus the best stocks for Apple grafting and Budding to have these characteristics in the future trees.

3. **Clone Perpetuation :** In some of the species, ordinarily is not possible to perpetuate the clones from vegetative cuttings, because the percentage of cuttings that root successfully is highly poor. But when the same cutting is perpetuated through Grafting or Budding has a very high success rate. Examples are that of Picea species in the forest tree's.
4. **Plant Damage and Repair :** The process leads to the repair of the old tree's and those damaged due to management, disease or rodents in the orchard are revived and rejuvenated through the art of Bridge grafting.
5. **Growth Rate :** Seedlings in case of certain fruit plants if let to grow naturally takes longer period for maturity. But when the same seedlings are propagated on a certain rootstock grow faster and add to an advantage of 4-5 years in maturity due to faster growth. Jackfruit is one such critical example.
6. **Indexing Virus** : Certain tree's which do not exhibit the disease normally are grafted or budded over a root stock to exhibit the disease or virus.
7. **Incompatibility** : Scion and rootstock when are incompatible, in such cases an Interstock is taken to bridge the compatibility and the problem is overcome.

Types of Grafts: Cleft Graft (Top working), Bark Graft (Change of variety), Side-Veneer Graft, Splice Graft, Whip and Tongue Graft, Saddle Graft, Bridge Graft (Repair of damage), Inarch Graft (Repair damage).

Budding : T-Budding or Inverted T-Budding, Chip Budding

Tools and Materials For Grafting, Budding: Budding and Grafting knives, polythene or polypropylene strips ½-1 inch wide binding strips of the grafts and buds.

Preparation for Budding : Prior to budding cuts are given in the rootstock with the knife. Prepared bud is then inserted in the cut and tight placement is secured as if it is natural. The bud is then secured properly by the poly strip so as not to allow any air contact with the operated parts.

Root Stock Production : Root stock is prepared from the seeds. Seeds first require a conditioning by way of rest period in their life for the embryo. This is achieved by seed storage under cool temperatures above freezing, moist condition. The seeds collection and procurement should be from certified sources to maintain the quality of the desired rootstock for a particular fruit. This maintains the uniformity and true representation of the characters desired. The seeds thus procured are subjected to other treatments prior to germination. These are stratification and soaking of the seeds and direct sowing to get highest germination percentage and avoid losses. Growth of the plant depends on the type of the root stock. In the case of Apples varieties grafted on seedling rootstock produce largest tree's and those grafted on clonal rootstock vary in size depending on the type of the clone.

Stock and Scion

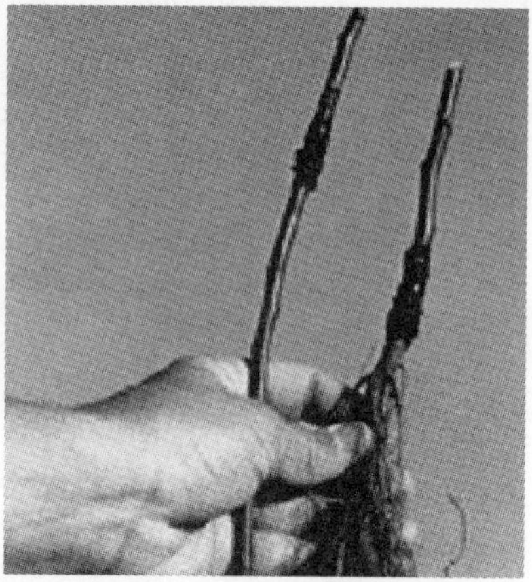

Grafted Plant

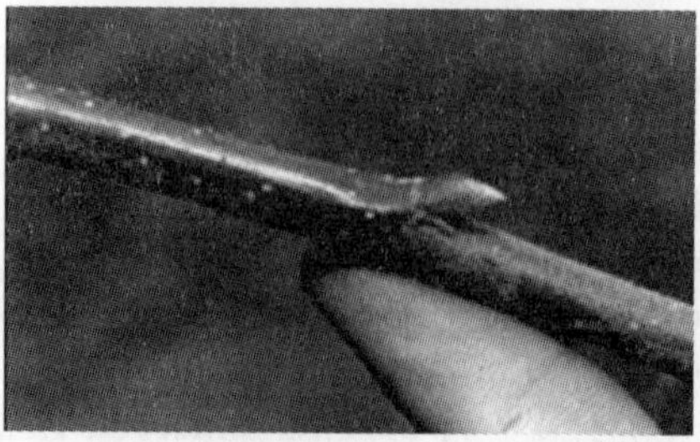

Selected Bud

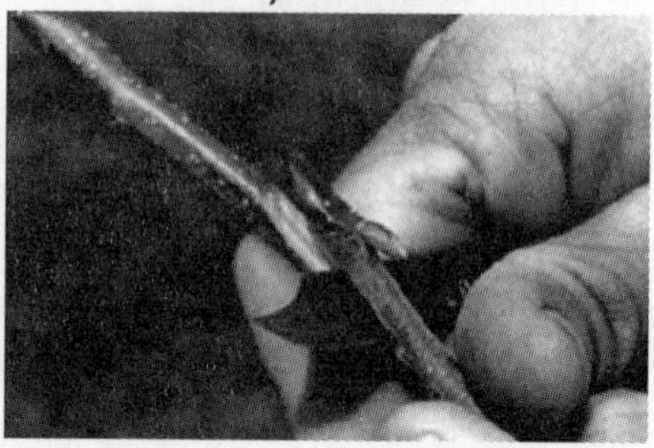

Bud Taking out

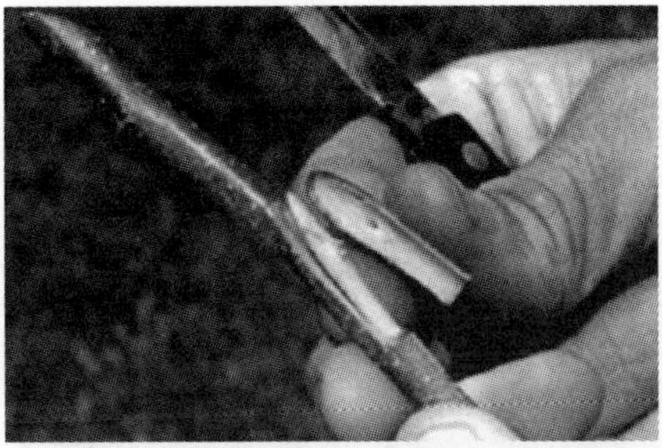

Bud

Bark grafting

Tying of Top worked stem

Top working

Bud Insertion

Bud Fixed & Tied

Bridge Grafting

Root Cutting

Stratification : Seeds which have hard seed coats need some external treatments to soften the seed coat and make them germinate. Different methods employed under this are Cracking or Chipping of seed coat by mechanical means, Abrasion, soaking in water, Scarification in acid and Stratification in sand or other moist media. In the stratification seeds are soaked for a certain period varying from 24-48 hours in water and placed in a growing medium of moist sand or peat moss or mix of both, till these start germinating. Plastic containers or wooden box's are the best for this process. Sand and peat mix is placed at the bottom nearly 2 inch thick and the seeds are then placed on this. Then another layer of Sand and peat mix 1 inch thick and again seeds are spread. The process is repeated for Nuts and other hard seeds but not for smaller seeds. These seeds start germinating after few months, which may range from 60-120 days depending on the type of the fruit plant seed. However the time period is 60-80 days for Apples, 45-90 days for Pears, 90-120 days for Peach, Plum and Persimon, Nuts. when the temperatures start rising. Germinated plants are then immediately placed in the beds at desired spacing for further growth. Care is very necessary for the germinated seeds not to damage the sprouts. In the case of other seeds these are soaked and planted in the seed bed directly. Soaking shortens the germination period. Discard stunted or diseased seedlings while transplanting.

Division : Suckers, Runners, Stolons of certain category of plants which have a typical type of fixing system in soil in them i.e. MINT, DOOB grass are multiplied through division. Sowing of Mint is done by collecting the runners during winters and then planting in spring. Doob grass is similarly made into small plugs, then planted, which then covers the whole area of the ground.

Layering : Numerous shrubs and plants are propagated with this system. Well known plants propagated with this system are Mango, Litchi, Figs, Berries, Carnations, Honeysuckle, Jasmin, Clematis etc. Layering is thus undertaken not only in fruits but also in ornamentals too. Layering is of two types i.e. AIR Layering and SOIL Layering. Air layering is undertaken in the case of erect plants like fruits which can not be bent downwards. This **Air layering** is also know as **Guttie**. Soil layering is done in the case of fleshy and ornamental plants which are bent down in the soil and covered with soil. After the rooting is complete the plants are severed.

Bulbs/Corms/Tubers/Rhizomes : These are the underground parts of the plant which are in one of the above forms. Known examples are the Onion, Garlic in the Bulbs category, Zaminkand as Corm, Potato and Sweet Potato in the Tubers group, Ginger and Haldi are the Rhizomes for propagation.

Tissue Culture : This is a laboratory process which takes care of working with a tissue of the plant for multiplication under perfectly controlled conditions. The multiplication of the tissue in millions brings out the plantlets of the system for further planting. These plantlets are fully acclimatized before planting in the soil.

Nursery Management: Plant material to be multiplied or propagated, needs the best of the care in the form of Planting media/Rooting media, plant care both outdoors and under covers. Provision of facilities like Misting, Fogging, sprinkler irrigation, Hormonal application techniques, fertilization. For commercial aspects of market sale of the multiplied materials requires special skills and techniques of creation and presentation of products for the purchaser or the consumer. All such care leading for a successful venture is the nursery management.

(a) Location of the Nursery with all facilities of water and Electricity

(b) Nearness to the Trading Center and easy availability of Skilled and Un-skilled labour

(c) Knowledge of Plant Propagation

(d) Type of Plant propagation whether for Horticultural Fruit Plants / Ornamental Plants / Arboriculture.
(e) Knowledge of Organic manure production and Plant protection
(f) Preparation of Potting medias.
(g) Skillful knack in the art of presentation and sale marketing
(h) Creation of Arte-facts from wood / Plaster of Paris / Clay
(i) Hormonal application technique.
(j) Management of Green House culture.

Kentia Palm Propagation : *Howea forsteriana* or Kentia palm is the queen of Indoor plants and is continent specific to Australia only. Annual exports to Netherlands is 9 million seeds and plants alone which earns Dollars for the country and has become a source of income generation for a tiny island in the eastern coast inhabiting 700-800 families. Export of Howea seed has been stopped since 1980 and only plant seedlings of the sizes 7, 10, 12, 15 cm are exported (Helen Moody-Floraculture International, April, 2001). It takes around 8 months for this plant to grow to the seedling export size. Production and export has been confined to the parameters of ISO-9002. The season lasts from February to November and the peak season is June-August. Germination percentage of the seeds is very low but with experience and trials it has been standardized at a level of 75% for trade of this special palm. Seed is germinated in sterile 50 : 50 peat-perlite mix inside sealed growing chambers that protect them from pests and diseases. Yield of seedlings is being continuously monitored for germination accelerants, temperature regimes at different stages of growth, and also extending the period to spread the growth of the seedlings. Seedlings are pulled, washed and are made ready for shipping by giving fungal treatments and then packing in sealed containers. Such like specialized propagating techniques are a must for the Indian farmer, Nursery manager to be adopted in the course of propagation and getting into the trade. It is this product quality which has niched a committed market for these palms.

Cuttings and Rooting Hormones : Hormones are an inseparable process in the growth of a tissue. In the process of making cuttings to root properly and bring up a clone exactly similar to the parent, rooting hormones are applied. Reason of applying rooting hormones to stem cuttings are :—

- To make rooting possible when too little rooting hormone exists.
- Increase the percentage of rooting.
- Increase the number of roots per cutting.
- To make cuttings to root more rapidly and thereby shorten the length of time and to keep the cuttings alive till it can support itself.

Rooting hormones applied in this job are synthetic very much similar in action to the naturally occurring hormonal compounds called Auxins. The two such synthetic hormones for rooting are Indole Butyric Acid (IBA) and Napthalene Acetic Acid (NAA). These hormones are required by the plant in very small quantities, as such these are available in the market under the very names in concentrated form. These are ready to mix formulations either with Talc powder or in solution form with water or Alcohol. High hormonal concentrations are as bad as not using a hormone. Stem tissue may get killed. Too high concentrations may withhold breaking of buds and making the cutting not to grow and root, thus killing the cutting in process rooting hormones are applied always to the fresh cutting. Rooting hormones are readily absorbed before plants begin to heal over the cut forming the cork tissue. Powders are usually best used as they come from the manufacturer. Liquid formulations are in concentrated form, that needs dilution. Label on the liquid formulation lists the active ingredient concentration which ranges from 0.5-2.0%. 1% range is equivalent to 10,000 ppm. This can be diluted as per requirement by adding water/alcohol. The cuttings in the solution are given only a quick dip. Treated cuttings thus be planted immediately in the nursery bed for rooting. As a thumb rule cuttings are dipped in the solution only ¼ - ¾ of an inch of the cutting, in the rooting hormone. Cuttings are then planted in the rooting media, keeping in view that the cuttings get self support. Thus the cutting should be generally put ½ - ¾ of the total length in the rooting media. In the propagation of cuttings a lot of care is given to the sanitation conditions, to maintain a good percentage of rooting success lest the cuttings are attacked by fungi and bacteria.

IV. ORNAMENTAL PLANT NURSERY

Gestation Period		2 years
		Cost in Rupees "000"
PROJECT DETAILS :		
Land	2.5 acres	Nil
Buildings	50 sq. Mtrs.	30
Green House 7 X 30 sq. mtr	200 sq. Mtrs	15
Tools and Implements – Knives, Secateurs, Saw, Spades, Fibbler, Pick axes, Sprinkler, Sporayer, Buckets, Baskets, Basins, Khurpas, Sickles etc.		15
Plant Material from New Delhi, Saharanpur and Local collection and fixing (Annexure – III)		45
Sinking of Tubewell	1 No.	30
Recurring Costs :		
Land preparation		10
Seed sowing, germination of various plant species and Raising of plant cuttings for rooting		10
Plant thinning and fixing in beds		5
FYM, Compost, Moss grass and other potting material purchase		5
Plant maintenance for 1 year		5
Purchase of Terracotta pots, Crafted pots, Cement pots, Poly bags, Moss sticks, Bamboos, Nylon ropes etc.		20
Man Power :		
• Trained Malee @ Rs 3000 per month	1 No.	36
Production		
• Rooted Cuttings, 1.00 Lakh @ Rs 2 per cutting		200
• Grown up seedlings and Plants, 1 Lakh No's. @ Rs 5 Per plant.		500
A. CAPITAL COSTS :		
• Land		NIL

• Building and Civil Construction	30
• Plant and Machinery	95
• Miscellaneous Fixed Assets	35
• Pre-operative Expenses	50
Total Fixed Capital	210
• Working Capital Margin	162
Total Project Cost :	372

B. MEANS OF FINANCE

• Promoters own Share capital	93
• Term Loan Finance	279

C. ANNUAL OPERATING EXPENSES

	Total Costs	Fixed Costs	Variable Costs
Variable Expenses:			
• Raw Material	45	-	45
• Packaging Materials	25	-	25
• Marketing and Sale Expenses	30	-	30
Total	**100**	**-**	**100**
Fixed Expenses :			
• Utilities	25	15	10
• Production Man power	66	36	30
• Farm Overheads	20	20	-
• Administrative and Management Expense	96	96	-
• Financial Expenses			
a. Interest on Term Loan	50	50	-
b. Interest on Working Capital	29	-	29
• Depreciation on Rs 95,000 capital cost	19	19	-
Total	305	236	69
TOTAL (Variable costs + Fixed costs)	405		169

D. NET SALE REALISATION

• Sales Realisation	700
• Contribution	305
• Pre-Tax Profit	69
• Gross Cash Accrual	88
• Surplus	69
• Total Cost	405

E. OTHER PARAMETERS

• Yearly Installment	47
• No of Installments	6
• Prime Lending Rate chargeable by Financial Institution	18%
• Rate of Interest on Working Capital	18%

ANNEXURE –IV

SHOWING ORNAMENTAL PLANT SPECIES FOR MULTIPLICATION

Plant Group **Name of the Plant**

Tree's and Shrubs

- Crotons, Coleus, Bauhenia spp. (Kachnar), Cassia spp., Jacaranda, Maulsari, Lagerstroemia, Callistemon (Bottle Brush), Graevillea, Kadamb, Ashok, Champa, Karonda, Duranta, Cupressus, Ficus spp.

Perennials

- Ferns, Palms, Orchids, Cycads

Foliage

- Draecena, Dieffenbachia, Fittonia, Agaloenema

Climbers

- Allamanda, Aristolachia, Beaumontia, Bignonia, Clerodendron, Ipomea, Jasmin, Tecoma, Lonicera, Passiflora, Bougainvilleas, Ficus repens, Hedera helix, Philodendron, Scindapsis, Vernonia

Bulbous Foliage

- Allocasia, Anthurium, Caladium, Maranta, Monstera

Bulbous Flowers

- Gloriosa, Hemerocallis, Hedychium, Hippeastrum, Hymenocallis, Tuberose, Gladiolus, Dahlia, Freesia, Canna, Nargis

Roses

- Polyanthas, Climbers, Ramblers, Miniatures, Floribunda, Hybrid Tea

Ground Covers

- Doob grass, Mexican, Zoysia

TISSUE CULTURE

Tissue culture plants and industry is a much needed future industry for India and its vast future as an export market earner viewing India a country of small industry, vast educated skills, more developmental attitude and growing need for providing employment avenues. In the light of all this establishment of a Mini Tissue culture Laboratory for the refinement of Protocols, testing of growth conditions (Chemicals, Light, Humidity, Temperature), disease indexing, breeding etc. all find an answer in Plant Tissue culture.

Project : To develop genetically identical, physiologically uniform plants at low cost through micro-propagation.

Objectives

- Cater to increased plant demand from the Departments of Horticulture, Forests, Industry (Mulberry for silk, Casuarinas for energy).
- To cater to creation of village employment opportunities through production, plantation raising and entrepreneurship.
- Manual Vegetative propagation technique qualifies for only smaller production part and to cater to, plant production in millions.
- Develop own techniques in breeding and development of plants with a special reference to pest and disease resistance.

- Developmental of Plant multiplication protocols for Horticultural, Forest, Vegetable and Herb plant species much in demand in the market.
- Export marketing of cut flower plantlets of Roses, Gerbera, Chrysanthemums, Orchids etc.

Process Methodology : Micro-propagation will be strictly followed as per protocol and know-how available for a particular plant. Tissue culture is to be initiated from any plant part and further to be induced to form shoots, roots and individual plants. Any of the processes for Somatic embryo genesis/ clonal / callus tissue etc. will be undertaken. After receiving the tissue from the mother plant the process starts as :—

- Excised tissue from the mother plant is surface sterilized, dissected into plant cells, placed in nutrient media under controlled environmental conditions (regulating light, temperature, humidity). acclimatized tissue grows into a shoot let.
- In the second step media composition is altered for development of mini-shoots to further develop real longer shoots.
- In this stage development of roots takes place, when roots develop on the growing shoots.
- In this stage the developed plantlets with shoots and roots are removed from the glass containers and are washed of nutrients and any sticking material. Plantlets are placed in a sterile rooting media (Peat moss or Sand and compost) in poly bags. Acclimatization under controlled conditions in a Green House is ensured through high humidity, low light intensity and transferring the materials from the lower to the high temperature conditions at desired intervals of time and bringing the plantlet to the natural conditions. The process is called plant hardening.

PROJECT DETAILS

- **Buildings :** Tissue culture Laboratory has a requirement for a growth and culture room, Media preparation room, store for glassware, chemicals, washing chamber, Distillation room.

- **Equipments :** Laminar Flow, Air conditioner's, Micro balance, pH meter, Incubator, Distillation still, Boiler, Glasswares, Chemicals, Microscope, Green House for hardening of plants and an open area for the culture of hardened plants.

Capacity : 4 million plantlets / year, with capacity utilization of 50%, 60%, 75% 90 capacity from 1st - 4th year.

PROJECT COST	**Rupees in "000"**
• LAND	100
• BUILDINGS : 80 sq. mtr area for Lab. Rooms @ Rs 5000 per sq. mtr accommodating Growth, culture room, media preparation room, store for glassware and chemicals, washing room, Boiler room, Genset room, Electrical control room etc.	400
• Green House 1000 sq. mtr.@ Rs 700 per sq. mtr.	700
• PLANT AND MACINERY : Incubator, Laminar Flows, Distillation unit, Boiler, Air conditioners, Tube-well, Autoclave	500
• UTILITIES : Diesel Genset, Airconditioning, Water	500
• MISC. FIXED ASSTES : Computers, Office Furniture, Transport.	300
• PRE OPERATIVE EXPENSES: Deposits, Travel, Insurance, Product development, Technical Know-how Fee's.	500
• RAW MATERIAL : Estimated 18% of the Total Sales	180
• MANPOWER AND SALARIES : Laboratory Manager, Production Incharge, Green House Manager, Daily wage workers 10 No's.	695
• PACKAGING MATERIAL	200

Estimate of Working Capital

Particulars	Period in Days	Amount required
Raw Material	90	180
Stores	60	150

Work in Process	90	500
Working Expenses	30	695
TOTAL		1525
Margin Money		382

PRODUCTION AND SALE

A. Capital Costs

• Land	100
• Building and Civil Construction	1100
• Plant and Machinery	500
• Miscellaneous Fixed Assets	300
• Pro-operative Expenses	500
Total Fixed Capital	2500
• Working Capital Margin	1000
Total Project Cost	3500

B. Means of Finance

• Promoters own Share capital	900
• Term Loan Finance	2700

C. Annual Operating Expenses

	Total Costs	Fixed Costs	Variable Costs
Variable Expenses			
• Raw Material	720	-	720
• Packaging Materials	200	-	200
• Marketing and Sale expenses	200	-	200
Total	1120	-	1120
Fixed Expenses			
• Utilities	100	30	70
• Production Man Power	300	210	90
• Laboratory Overheads	50	50	-
• Administrative and Management Expenses	395	395	-

• Financial Expenses			
A. Interest on Term Loan	504	504	-
B. Interest on Working Capital	180	-	180
• Depreciation	350	350	-
Total	1879	1539	240
Total cost (Variable + Fixed)	2999		1360

D. Net Sale Realisation

• Sales Realisation	2000
• Contribution	1879
• Pre-Tax Profit	240
• Gross Cash Accrual	370
• Surplus	240
• Total Cost	2999

E. Other Parameters

• Yearly Installment	400
• No of Installments	7
• Prime Lending Rate chargeable by Financial Institution.	18%
• Rate of Interest on Working Capital	18%

CHAPTER - 8

CROP SPECIFIC POST HARVEST TREATMENTS

Post harvest requirements for the cut flower industry are the shelf life of the flower with the end consumer. It is this quality which is the main value addition for the floricultural products whether cut or potted. In the case of cut flowers the flowers are very delicate and amenable to every type of damage right from handling to storage and transport to the auction table and the retail shop. Ethylene though is a growth hormone but plays its role for senescence of the flower stem. All cut flowers has thus been categorized as ethylene sensitive and non-sensitive to ethylene. Fungal and bacterial attacks clog the stem and block the translocation of food and water and this leads to very many problems in quality degradation but also in pre-mature senescence of the stem. Thus post harvest treatments are crop specific for every individual flower.

AGAPANTHUS

Common Name : Lily of the Nile, Blue African Lily

Agapanthus are Ethylene sensitive flowers. For extending life, treat flowers for one hour at room temperature or for over night under refrigeration Then place the flowers into a holding solution with STS (Silver thiosulfate) or any proprietory chemical like Floralife STS - Silflor. Holding solution is a fresh flower food. Foods are available in the market like Floralife Crystal Clear. Troubleshooting for Agapanthus can be due to excessive drying out, rough handling or exposure to Ethylene.

ALSTROEMERIA

Common Name : Peruvian Lily Alstroemeria are Ethylene sensitive flowers. These benefit much from STS treatment. Treat flowers at room temperature for one hour with STS. At the site of arrival with the wholesaler or retailer, advise re-cutting of stems preferably in water. Then place the flowers in holding solution, which is a fresh flower food. Proprietory foods like Floralife Crystal Clear, which provides energy in the form of Carbohydrates, maintains uptake of water and inhibits stem plugging for proper flow of nutrients and water. Troubleshooting problems in Alstroemeria are yellowing of foliage, buds fail to open, abscission or flowers fall. Storage temperature of 1-2°C with 80-90% RH levels is recommended.

Alstroemeria

ANTHURIUM

Common Name : Flamingo Lily or Oilcloth flower Anthuriums are not Ethylene sensitive. Treat cutflowers for hydration especially when Anthuriums are to be shipped as dry pack after harvest. Treatment time varies with temperature and storage conditions. At ambient room temperature time can be ½-1 hour. Proprietory Hydration chemicals encourage water uptake. Hydraflor/100 is one such product which is in concentrated form and Hydraflor Quick Dip is a ready to use product and requires no mixing or dilution. Recutting of the stems is a must under water at the market site. Holding solution for increase in shelf life is a must of the post harvest process. Anthuriums are stored at 8-10°C with 80-90% RH level in the atmosphere. Anthuriums loose gloss quickly which is caused by prolonged exposure to high temperature and water stress. Blueing or blackening of the cut Anthurium stems is due to chilling injury.

Anthurium

CARNATION

Common Name : Carnation

Carnations and miniature Carnations are Ethylene sensitive. Treatment with STS is highly beneficial for Carnations. Treat Carnation flowers for one hour with Silflor at room temperature. Then place the treated flowers in a holding solution (Floralife Crystal Clear) and hold flowers for 1-2 hours in this solution. Advise re-cutting of stems by the wholesalers and retailers at the time of receipt with them under water. Store cut Carnations at 1-2°C with 80-90% Relative humidity levels. Problems which may arise are split calyx, buds failing to open, sleepiness because of exposure to Ethylene or high temperatures during shipping and storage. Try to store and ship Carnations alone and not with any Ethylene producing product.

DAFFODIL

Common Name : Narcissus

Daffodils or Narcissus require hydration or holding solution treatment for longer life of the cut flower. Advise immediate un-packing and re-cutting of the stems and putting in a holding solution. Proprietory products available in the market for holding solutions are either in concentrated or ready to use form. Farmer producer can use any of these depending on his convenience. Then store treated cut flowers at 1-2°C with 80-90% RH level in the store.

DELPHINIUM

Common Name : Larkspur

Larkspur is highly Ethylene sensitive flower as such must be treated with STS. at room temperature and the put into holding solution which provides food and energy to the cut flower stem. Hold cut flowers for 2-3 hours in the solution. Problems arising are the shattering of the florets, which can be due to drying or water stress, rough handling, old age or exposure to Ethylene.

FREESIA

Freesias are ethylene sensitive cut flowers. Flowers colour may fade due to ethylene exposure or high temperatures. Florets shatter and remaining florets do not open are the common problems encountered in the marketing of Freesias. Usage of STS treatment and hydration treatments are a must for long life and maintaining freshness of cutflower stems.

GERBERA

Common Name : Gerbera, Transval Daisy

Gerbera is non sensitive to ethylene, but requires definite hydration for supplementation of nutrients for cut life increase for 1-2 hours at room temperature. Store cut flowers at 1-3°C. Application of Hydraflor a proprietory product induces the strength required for storage, vase life. Water stress and fluoride content in water may cause limpness and necrotic or brown petal tip appearance.

GYPSOPHYLLA

Common Name : Gypsophylla

Gypsophylla is ethylene sensitive flower and requires STS treatment immediately after harvest. Standard timing for the process is placing of the cut flower stems in STS or STS compounds of proprietory origin for an hour to 1-½ hour. Then treating the stems for hydration. Exposure to ethylene or disease development or age of the cut stem leads to excessive browning of flowers. Dry conditions or water stress to the stem may also lead to non opening of flowers. STS and Hydration treatments save the cut flower from these problems.

IRIS

Iris flowers are non sensitive to ethylene, but do need Hydration for ½ an hour immediately after harvest and then subjecting the flowers to nutritive food application or subject to holding solutions. Drying out of flowers and foliage, buds failing to open and yellowing of foliage are the other problems of routine which can be due to improper storage or insufficient solution uptake etc.

ROSE

Rose is the topmost flower of the cut flowers in demand and deserves the best of attention and treatments and care in its storage, post harvest handling and dry packaging so as to ensure fresh delivery at the auction center. Harvested Rose stems are immediately put into buckets containing water for hydration. Use of special hydration solutions is recommended. Treatment time depends on temperature and storage conditions. Rose is sensitive to ethylene resort to STS treatment if it has to be transported with ethylene producing materials depending on the variety. Ethylene sensitivity varies with varieties. During storage place cut stems in holding solution for 1-2 hours to make certain that in dry pack these arrive in best condition at the auction center. Failing of buds to open, pre-mature bent neck, frequent petal burning are the causes of imperfect post harvest processing which is phase of extra stress on the cut flower.

Rose

STATICE

Common Name : Statice or Limonium

Statice after harvest is immediately put to a quick dip in Hydraflor before placing into preservative. Flowers are then put in the holding solution. Improper storage conditions or bad cultural management may lead to excessive yellow foliage and mold development due to excessive water present in storage and shipping.

TULIP

Cut flower stems of Tulips immediately after harvest are put in water containing Hydraflor for ½ an hour. Then the flowers are transferred into a properly mixed solution of Floralife Fresh Flower food. Store flowers vertically at 0-1°C. At the receiving end advise for storage of cut flowers in a holding solution and keeping this for 1-2 hours. Tulip flower stems tend to grow even after harvest. This is because of high temperature above 2°C. Flowers has the tendency to open quickly at these warm conditions.

Lilium

LILIUM

Cut stems rather than pulling. Harvest Lilium stems in the morning to prevent drying out. Cut stems at proper maturity stage. Harvest stems with 10 + buds when there are at least 3 coloured buds, stems with 5-10 buds when there are at least 2 coloured buds, stems with less than 5 buds should be cut when one bud is coloured. After harvest cool stems at 1-2°C as quickly as possible. Bunch stems only after cooling. While bunching remove leaves from the lower 10 cm. Store stems upright in clean water in the cooler to hydrate. Pack stems in perforated box's with holes to prevent ethylene build up. Provide shipping temperature of 1-2°C. For long term shipping pre-cool box's prior to loading. At the auction center retailer or wholesaler should re-cut the stems under water before placing these in vases with floral preservatives.

CHAPTER – 9

FLOWER CULTURE

FLOWER CULTURE

Floriculture is both a part of enjoyment and a profession for earning from the land. Flowers exude the natural happiness of joy and closeness with nature through their colour, size, fragrance. In the home garden flowers in the herbaceous borders with size and colour is a look of serenity and soberness, beauty and landscape designing. Culture of flowers depends on full knowledge about the seasons, species, flower habits and agronomical requirements for productive units. Annuals raised from seeds are the most important flowers for the home gardens. For the professional annual flowers should be good in market demand and as cut flowers. Annual flowers suited for Indian conditions are :—

Table : Showing Annual Flowers, Physical characteristics

Flower	Days for Blooming	Plant height in cms	Spacing in cms	Characteristics
WINTER ANNUALS				
Ageratum	55-70	25	25	Blue, Crimson and pale salmon pink colours

Alyssum	45-55	10-15	10-15	White, Muave-pink, Lilac
Anchusa	80-85	40-50	25-30	Blue
Antirrhinum	90-105	15-22	10-15	Various colours : white yellow, pink, maroon, orange, apricot
	45-60	25-30		
	90-100	25-30		
Brachycome	75-85	15-20	15-20	Lilac, Lilac blue, White, Pale blue
Calendula	55-70	25-30	20-30	Orange, yellow, Lemon, Off white in single and double sizes
Candytuft	55-70	30-45	20-25	White, Lilac, Carmine
Cineraria	70-85	35-45	25-30	Blue and Magenta, Bicolours with white blue, white magenta
Clarkia	75-80	90-100	40-45	Pink and Salmon shades
Cock's Comb	70-80	15-20	15-20	Yellow, orange, maroon, bright maroon, red, shades of cream, yellow, pink, orange
	60-75	35-40		
Coreopsis	55-70	30-45	25-30	Golden in perennial, many ranges in annual
	60-90	25-30		
Cornflower	65-70	60-75	40-45	Blue, pink, white, purple-blue in both single and double.
Delphinium	95-110	45-60	25-30	Blue with pale center and more shades in blue lilac.
Dimorphotheca	80-85	45-60	25-30	Orange, yellow, apricot lemon and shiny white
Godetia	80-85	45-60	30-45	White, pink, mauve, crimson.
Gypsophylla	65-70	55-60	25-30	White, pink
Helichrysum	80-85	90-100	40-45	Golden bronze, pink, brown, red.
Hollyhock	100-110	150-250	60-90	Pink, red, yellow, cream, white, maroon, in single and doubles.
Lady's Lace	70-75	90-120	45-60	White
Larkspur	80-85	80-100	30-35	Muave, Purple, Pink, Salmon white, Red. Single and Double
Lupin	80-85	60-75	45-60	Blue, Light pink European cultivars in bicolours, with yellow, mauve, white-red, white-blue, yellow-blue.
Mesembryanthum	82-90	10-15	15-20	Magenta with white center, golden yellow, brown, scarlet-red, cream with crimson.
Nasturtium	75-85	20-30	20-25	Yellow, orange, maroon, scarlet. Single and double. Bicolours also.

Nemesia	85-100	30-40	15-20	White, yellow, orange, pink, mauve, bronze red.
Pansy	105-115	10-15	15	Single and double with very many colour hues and colour sprays of petals
Petunia	110-130	20-40	30-40	White, red, salmon, scarlet, mauve, pink, purple.
Phlox	100-115	20-30	25-30	Both annuals and perennials with presentation of all colour hues, shades. White, pink, yellow, crimson lavender, mauve, pink, lilac.
Statice	105-115	30-45	25-30	Rose pink, blue-muave, yellow.
SUMMER ANNUALS				
Amaranthus	80-85	50-100	50-70	
Aster	80-85	20-25	15-25	Various colours and shades of white, cream, pink, grey, blue, mauve, maroon. Single & doubles.
Balsam	40-45	45-50	20-25	White, purple, rose pink, red, mauve with different shades. Single and double
Coleus	80-85	45-60	25-30	Foliage plant with wide range of colour combinations of dark, pale green, brown, purple, cream leaves.
Cosmos	50-55	100-150	40-45	White, pink-rose, mauve, crimson. Single and double.
Galliardia	80-85	30-45	45-60	Single and double flowers. Yellow, brown, red colours.
Gomphrena	70-85	40-45	25-30	Purple, orange, pink, off white.
Marigold	55-60 80-85	15-20 40-100	15-20 60-65	Yellow, orange, Lemon, dark brown and combination of these and other shades. Single and double. New mutant in white is also available now from USA.
Portulaca	55-70	15-20	10-15	Yellow, scarlet, crimson, rose, pink, magenta
Zinnia	40-55	15-30	20-30	White, dark brown, maroon, copper red, yellow. Available in single and double. Doubles compact and decorative.

ROSES

Rose is the oldest flower in culture which has the highest consumer acceptance in the world because of very many varieties, colours, forms, types, fragrance, growth habits. Roses account for 50-60% sales in the Dutch markets and 39% in USA. Roses have been classified in 5 categories. Hybrid T's, Floribunda's, Polyanthas, Miniatures, Climbers. This classification is based on the present day requirements of show and modern Rose culture. Hybrid Perpetuals are hardy.

Hybrid Teas : Hybrid T's are tender, need protection and possess continuous flowering habit. Flower are large and originated from cross between Hybrid Perpetuals and the Tea Roses. They are bushy and vary in height and other characters. Varieties are highly excellent in Hybrid Teas with regard to single colour / bicolour.

Hybrid Tea

Floribundas : Popularity of Floribundas is for producing large quantity of blooms more than Hybrid T's. Floribundas are hybrid Polyanthas. Floribundas outcome of the industry owes its credit to Danish Breeder Poulsen during 1910. This is a cross between Hybrid T's and Polyanthas. The plants are large enough like Hybrid Teas in form but the blooming is the character of dwarf Polyanthas with a smaller flower to Hybrid T.

Polyanthas : Polyanthas (Pompons) or Dwarf polyanthas has a unique characteristic that these bloom for months together, produce flowers in profusion either in trusses or in a mass of single or double blooms and are much smaller in size to Floribundas and Hybrid T's.

Climbers and Ramblers : Distinctive class of climbers is known as Ramblers. These bloom freely in hills only. Climbers are sports of Hybrid Teas or crosses with Ramblers. Ramblers are best adopted for Landscaping of arches, pergolas. The best Indian varieties in this category are Climbing Peace, Climbing Virgo, Pink Mardan, and Delhi Pink Pearl.

Miniatures : These are the baby or miniature Roses and are much popular in the Landscape and flower arrangements. These are used for edgings, in Rock gardens. Famous varieties in the group are Cri-cri, Baby Masquerde, Little Sunset, Rosmarin, Starina.

Essentials for Rose Culture

- Require open sunny culture without shade.
- Soil should be rich and well drained.
- Love optimum moisture condition in the substrate.
- For productive crop culture soil should be rich in both macro and micro nutrients.
- Hybrid T's culture should be from selected rootstock.
- Miniature culture should be on their own rootstock to maintain dwarfness.
- Avoid low pH for productive Rose culture.
- Select disease resistant varieties.
- For exhibition and shows resort to disbudding.
- Trim, remove suckers and dead wood from the plants.

Rose Colours : Orange, Scarlet, Blue, Pink, Deep Pink, Red, Yellow, White, Bicolour and Multicolour are a few colour variations in the market of Rose culture.

Varieties : Indian varieties are much different from the European varieties. Some of the beautiful and highly in demand Indian Hybrid Tea varieties are Super Star, Montezuma, Michele Mieland, Papa Meiland, Crimson Glory, Granada. The European varieties are First Red, Red Berlin, Bianca, Red Serenade, Vivaldi, Lambada, Souvenir, Escimo, Rodeco, Sacha, Vanilla, Kiss, Ravel, Baronesse, Tantau's Konfetti, Prophyta, Frisco etc. Each shoot bears a flower at the tip.

Beautiful Indian Floribunda varieties are African Star, Banjaran, Himangani, Iceberg, Queen Elizabeth etc. European Rose varieties have been bred keeping in view the customer fancy for colour, stem length, vase life. Most demanding colours and Rose varieties which have the maximum demand are :-

RED

- **Forever Young :** 70-90 cm stem, huge flower head.
- **Solo :** 70-100 cm long stem, with medium large head.
- **Vermilion Red, Rouge Baiser :** 70-100 cm long stem
- **Valentino :** 13 day vase life, 240-260 stems per sq. meter production, with average length of 60 cm.
- **Salsa** : Bright Red, thornless plant, with medium head, 14-16 day vase life and a productivity of 300-340 stems per sq. meter.

PINK

- **Byblos :** Medium pink with 15 days vase life.
- **Fenice :**Pale pink, with 12 days vase life.
- **Swing :** 50-70 cm stem length, with 21 days vase life.
- **Sioux :** Salmon pink, 60 cm stem length, 18 days vase life, with a productivity of 200-300 stems per sq. meter.
- **Leandra :** Apricot pink, with large head, 14-16 days vase life.

MEDIUM YELLOW AND GOLD

- **Yellow Island :** 70 cm stem length, with 15 cm bloom.
- **Gold Strike :** 12-14 days vase life.

WHITE AND CREAM

- **Vadela :** Ivory white beauty lasts for 20 days.

Soil : Best soil for Rose culture is judged from the depth of the soil, texture, structure, drainage, availability of soluble nutrients and Rose growing pH level above 6.5.

Transplanting and After Care : Ensure the selected plants are free from root damage and the roots are moist enough to support the growing plant whether in open or covered culture. Plants are transferred and fixed at site immediately and watered.

Manuring and Fertilization : Indian soils are deficient in Nitrogen and Phosphorous and seldom in Potassium too. Soil analysis is a must to start Rose culture. Deficiency in any of these may restrict plant growth. Organic manure supplements of plant or animal origin in the form of composts, FYM, animal meals, oil cakes highly usefull in Rose culture. These are added either at the time of soil preparation for transplanting or at the time of pruning. Roses are known to suffer from deficiency of micronutrients. Deficiency symptoms are yellowing of leaves due to Nitrogen, Intervienal chlorosis with viens remaining green is due to Manganese, dying of terminal buds mark absence of Boron and centers of leaves becoming pale, developing of dead tissue around mid rib for Magnesium deficiency.

Preparation for cut flowers, starts from the 2nd or 3rd week of October immediately after pruning. Rose plants are dug upto a depth of 5-6 inches around the plant and are left as it is for 5-7 days. The process is done to solarize the soil for making it free from soil borne pests and microbes which may hinder in its culture later on. The plants are added with a dose of 3-4 kg of well rotten FYM or compost or vermicompost, which is mixed well with the dug up soil and plant is filled and placed in its original position. After 10-15 days the plants are side dressed with a blanced mixture of NPK fertilizers. Recommended application of fertilizer Foliar spray by the IARI is Urea, DAP, Potassium Nitrate, Potassium Phosphate in the ratio of 2 :1 : 1 : 1; or 30 gm of this mixture is dissolved in 10 liters of water and is sprayed. Micronutrients are applied in the form of Ferrous sulfate, Zinc sulfate, Manganese sulfate, Magnesium sulfate each 20 gm, Borax 5 gm. 20 gm of the mixture is then dissolved in 10 liters of water and is sprayed on the plants. At the time of flowering no spray is recommended.

Pruning : Pruning is a process of plant stimulation to make quick new vigorous growth, removal of dead, weak and twiggy, unwanted tissues. Pruning also regulates the size and shape of the plant. In India pruning is done immediately after the monsoons is over, but under Delhi conditions the best time judged for pruning of Roses is during October. However in Europe and Western countries of the temperate region, pruning is done immediately after the frost is over and spring is just to set in. Frequency of pruning is advised only once in a year. Pruning is well defined to be different for each category.

Propagation : Roses are propagated from seeds, cuttings, layers, budding and grafting. Breeders resort to propagation from seeds. Climbers, Ramblers are propagated from cuttings,

Layering. Grafting and Budding is done in case of Hybrid Teas and Floribundas, where the scion is transferred on to the known rootstock of good quality, which affords the growth of the plant. For this very many good stocks have been identified. Edouard Rose and Rosa multiflora are the best Indian rootstocks. In the western countries Rose stocks are from either of the Rosa species of multiflora/laxa/canina. In the plains of India Eduord Rose stock is the best. In other Indian states and in the hills Rosa multiflora stock is used for the propagation of Roses. Rose propagation is mostly preferred from budding technique rather than adopting other methods.

After care in propagation is one of the most important aspects. After budding the eye shows signs of swelling and top portion of the stock above the budding point is cut off after successful union of the eye with the tissue of the stock. Fiber or polythene or rubber band is then removed. Care is exercised to remove all the suckers as these show emergence from the stock below the union.

Spacing : Plant spacing in Roses is further dependent on the type of the Rose variety whether a Hybrid Tea or a Floribunda or a miniature or Climber. Density is the other consideration, which has a direct bearing on per sq. meter production. However this is to be adhered to on the recommendations for the variety. In covered crop culture 7 plants per sq. mtr is the density which is most convenient and productive.

Potting Soil : Substrate media for the Roses to be planted in pots or in improvised trenches for cultivation under covers, conforms to the qualities of successful Rose culture. Potting mix or Bedding or growing media is rich in its plant nutrients, well drained, porous with high organic content and very good moisture holding capacity. Various medias which are being used for making pot mix are Peat Moss / Sphagnum Peat Moss, Coco Peat, Vermiculite, Perlite, Sand soil, organic manures like vermiculture/ poultry / rhizobial cultures etc.

Pests and Diseases : Beetles, Scales, Aphids, Jassids, Leaf eating, Mites and Bugs are some the pests of the Roses. For a well managed commercial Rose culture control of pests like Aphids, mites, Jassids and thrips is of prime importance. Spray of contact and systemic pesticides is highly recommended. Pests which have been reported from other parts of India are Tussock caterpillar, cator slug, crab caterpillar which feed on leaves. Recommended pesticides are Diazinon, Aldrin, Malathion, Pyrethrins. In the diseases dieback, stem blight, powdery mildew, rust and viruses are the one's which cause considerable damage to the crop. Resistant varieties selection, covered crop culture and use of biopesticides are the best remedies to catch up with the market conditions for sale and export. For the control of these diseases properietory fungicides available in the Indian markets are wettable sulphur, Benomyl, Carbendazim, Captan, Ferbam, Zineb.

Production : Production of Rose cut flowers is based on the crop density per acre/hectare. Generally 7-10 plants are placed in one sq. mtr. and each plant has the capacity to produce 16-24 flowers under open culture system and up to 30 flowers, per plant under Green House or covered crop culture. Cut Flowers are then graded as per the length of the stem. These cut stems are placed in three grades A,B,C. A grade is the top grade and produces nearly 60% flowers with B grade following next and then C grade.

Rose culture is also being done for the production of Rose oil (Otto of Roses) and Rose water. Suitable varieties for oil and Rose water production are *Rosa damascena, R. centifolia, R. alba, R. gallica.* CIMAP, Lucknow has recommended only Rosa damascena for Otto and Rose water production. Material is available on demand from the Regional Research Laboratory Palampur, District Kangra (Himachal Pradesh).

ROSE CUT FLOWER COMMERCIAL CULTURE

Commercial culture of Rose crop as a cut flower in the open is very much popular in the Indian metros of Delhi, Calcutta, Mumbai and Chennai, where demand for cut Rose flowers for Bouqets, flower arrangements, presentation is too much. Growers in all these metros have gained with their experience and very much lucrative crop. In Delhi, to give a boost to Rose culture as cut flowers for domestic and export markets an auction center has come up, which further has guaranteed profitable returns to the growers. Soil is prepared by giving 2-3 turnings with a cultivator and disc harrow and planking. Beds are the raised in the soil and pits are dug. Pits are supplied with FYM and fertilizers and then the plants are placed in these pits. Process of preparation begins from the month of October with the procurement of plants from a certified Rose plant Nursery of the area. Plants start flowering from the month of February and continue till June during this period the crop is well managed for its growth by applying irrigations and giving proper hoeing and weeding operations. Cut flowers are harvested in the very early morning hours and are prepared for early market delivery and auction. Cultural cost per acre basis are as under.

CULTURAL COST

AREA :	1 acre
PLANT DENSITY :	8 plants/sq.mtr
• Irrigation of land prior to tillage	400
• Land preparation	1000
• Preparation of beds	500
• Digging of pits, 32000 pits @ 300 pits / man	8000
• Preparation of pits for planting –20 men	1500
• Planting of Rose plants @ Rs 8 per plant	320000
• Cost of FYM, deoiled cakes	7000
• Cost of Fertilizers NPK (100:120:100) kg per acre	1600
• Plant protection costs	5000
• Irrigation 15 –20 No's	4000
• Hoeing and weeding	5000
• Management Expenses	80000
• **Total:**	**370000**

Production : Total Cut Flowers – 650000 No's.

- **A** Grade - 3.2 Lakhs
- **B** Grade - 1.9 Lakhs
- **C** Grade - 0.65 Lakhs

Auction Returns : A grade @ Rs 12 per dozen, B Grade @ Rs 7 per dozen and C Grade @ Rs 5 per dozen. Total auction sale for Rs 4.9 Lakhs is estimated brings an estimated return of Rs 1.2 Lakhs. Investment on the cost of the plants in the first year is also a profit, in the form of return on investment in the second and subsequent years.

CARNATIONS

Carnations are beautiful garden flowers and command number three position in the world of cut flowers. Carnations belong to the genus Dianthus. Requirements of the Carnations are open sunny site with well drained rich soil. Carnations are perennials in the hilly regions but behave like winter annuals in the plains. Carnation flowers are grown in the world for cut flowers industry, ornamental show. Another fact which is not popular amongst the flower growers is the sweet aroma which it possesses and is used to a limited scale for the extraction of concrete, which can be converted into absolute by alcohol washing. Its concrete content varies from 0.2-0.3%. Carnation culture for the production of concrete, absolute is very limited in Germany, France, Switzerland and Italy. Aromatic concrete absolute is used to intensify notes of roses, lily and other floral notes. Main constituents contained in the Carnation flower are : Eugenol, Phenylethyl alcohol, benzyle benzoate, methyl salicylate.

Middle hill areas in India provide a very conducive climate for all round the year culture of Carnations. Carnation flowers are available in two very distinct types which are (1) Standard Type (2) Spray Types. Colours available in different cultivars in both the types are :—

Standard Type		Spray Type	
Colour	Cultivar	Colour	Cultivar
Red	Scania, William Sim	Red	Vermillion Protruding Peach Delight
White	White Sim	Purple	Equisite, Royalette
Pink	Lena, Dusty Pink	White	White Lilia
Crimson	Alec Sparkes	Pink	Sam's Pride
Yellow	Yellow Dusty, Tangerine Sim	Yellow	Goldi Locks
Variegated	Arthur Sim, G.J. Sim, Red Diamond	Variegated	Twinkle

Propagation: Carnation is propagated by cutting, layering, root division and seeds. For the perpetuation of the renowned mother stock which has already done well is preferred. Cuttings is the proven method of propagation for Carnations. However root division can also be done. The process is recommended for Indian plains, but in the case of hill culture, plants must be repotted, which can remain for 3 years. For Pinks root division and seed are preferred. Sweet William is generally raised by root division. Perpetual carnations can be kept into flowering through out the year in the hills by proper culture and disbudding.

Planting : Planting time of the propagated material whether by seeds or by cuttings is different climatic zones and areas. In the high hill areas of North, the planting is undertaken during March-April and in the plains of India, it is cultivated as a cool season or winter crop and sowing / plantings are done from September to November.

Spacing : A spacing of 40 x 40 cm is very much optimum for higher yield and good quality flowers. Spacing distances are different with growers in every area. YS Parmar University of Horticulture and Forestry recommends planting of rooted Carnation cuttings 1.2 meter wide raised beds at a spacing of 0.75 mtr or 2 ½ ft from plant to plant. In normal plantings raised from seeds spacing is generally adopted as 30 x 30 cm thereby accommodating 1 plant per sq. ft or 9 plants per sq. mtr. This gives a planting density of 43000 plants per acre.

Manuring and Fertilization : Carnations are very light feeders of nutrients and not much work has been done on the application of manures and fertilizers on this crop. Nitrogen is much important a factor for Carnations than Potash and Phosphorous.

Use of Growth Regulators: Use of growth promoting hormones is a very common practice for Carnation in promoting the branching and reducing the effective height of the plant. NAA at 500 ppm is used for branch promotion and Ethrel at 500-1000 ppm is used for reducing the plant height.

Disbudding / Pinching : This process is undertaken for the market demand for good flower size and production increase per plant. Disbudding prepares the stem for the formation of flower clusters for exhibition. The process is undertaken between 80-90 days after sowing. Disbudding of auxiliary buds is done after 7 days of the appearance of auxiliary buds for better quality flowers.

Pests and Diseases : Red spider, Aphids, Thrips are the major pests of Carnations. In the diseases wilt is most common.

Split Calyx : This is a defect in the culture of Carnation which appears in the flowers and reduces the quality and market value of the cut flower. This is due to wrong cultural practices or defective environment i.e. excessive heat or moisture or defect in the variety.

GLADIOLUS

Gladiolus is widely known the world over for its attractive cut flowers for flower arrangement and bouquets and as a bedding plant in the landscape. Gladiolus has very attractive colours and shades in it varying from white, cream, pink, yellow, scarlet, crimson, maroon, lilac, apricot, orange, rose, violet.

Improved Varieties: Very many present day Gladiolus varieties have been bred and developed in USA and European countries of Holland, UK. In the early seventies varieties from these countries were imported and evaluated for acclimatization and good market characters. Introduced varieties of the time were Anne Virginia, Cardinal spellman, Blaur Domino, Fenny Lind, Florence Nightingale, Hawaii, Hill crest, Oklahama, La Paloma, King Lear, Double Frills of pink etc. Hybridization work at Indian institute of Horticulture Research Hessarghatta, then started the evaluation and improvement of Gladiolus varieties for Indian culture. Varieties developed for India are Meera, Nazrana, Poonam, Sapna, Aarti, Apsara, Agni Rekha, Mayur, Suchitra, Manmohan, Manohar, Mukta, Manhar, Manisha, Mohini, Jawala, Archana, Arun, Sanyukta, Triloki. Development works for selection of more Gladiolus varieties were undertaken at Mashobra, Chaubatia, IARI and NBRI Lucknow. Varieties based on colour are:

- **Crimson** : Emelia, Majolia, Oscar, Jo Waggoner.
- **Red** : Atlantic, Albert Schewitzer, Hawaii.
- **Scarlet** : Ahoy, Dr. Fleming, Sansouci, Elen.
- **Orange** : Oscar, Tunia Elite, Pink.
- **Salmon** : Bon Voyage, Sunny Boy, Bloom Fontain.
- **Rose** : Friendship, Lovely Melody, Spring song, Spick and Span.
- **Blue** : Bluebird, Blue Conqueror, Blue Ganga, Dutch Beauty.
- **Purple** : Blue Boy, King of Blacks, Manju.
- **Yellow** : Garden Wattle, Happy End, Yellow Emperor, Polugoon.

- **White** : Eiffel Tower, Morning Kiss, Maria Goretti, Beverly.

GLADIOLUS VARIETAL DESCRIPTION

- **Aarti** : Developed from a cross between Shirley and Melody. Spikes are ready after 70 days, very attractive with Poppy red florets, 11 per spike. Vase life of the cut flower under ordinary conditions is 6 days.
- **Agni Rekha :** Open pollinated seedling of the cultivar Sylvia and a mid season maturing. Plants attain a height of 90 cm. Florets are fire red in colour with saffron yellow blotch and scarlet stripes, with 18 florets. Very suited for cut flower culture.
- **Apsara :** Variety developed by hybridizing Black Jack x Friendship. It flowers in 45 days after planting. Spikes are 97 cm long. 18 florets are there per spike with Ruby red florets. Vase life of the cut flower is 8 days under normal conditions of storage.
- **Archana :** A cross between Sylvia x Friendship takes 90-120 days for maturity. Spikes are 80 cm long with a branching habit. Florets are Bogonia with a central white streak on petals and blotch is Primrose yellow. 16-18 florets are their on the spike.
- **Arun :** A cross between Sylvia x Fancy requires 90-120 days for maturity. Spikes are 80 cm long and branched. Florets are Vermillion in colour.
- **Manmohan :** The hybrid a cross between Friendship x *Gladiolus tristis*, blooms after 80-120 days from planting. Spikes are 80 cm in length. Florets are Primrose yellow having irregular splashes of orchid purple at the tips of outer 3 petals, throat Primrose yellow having splashes of orchid purple. Florets are 14-16 per spike. Cormel production is very good.
- **Manohar :** This is again a cross developed from Friendship x *G. tristis*. This also flowers in 80-120 days from planting. Difference between Manmohan is that the florets of Manohar are Orchid purple. Spike length is 80 cm.
- **Meera :** This is a cross between G.P.1 x Friendship. It flowers after 58 days from planting. Spikes are robust and

attain a length of 90 cm. Variety is highly good for garden displays. and as cut flower.

- **Mukta :** This is the 3rd hybrid from cross between Friendship x *G. tristis*. Florets are sulphur yellow in colour and spike attain a length of 70 cm, takes 90-120 days to maturity.
- **Manhar :** Manhar is the 4th hybrid from the cross between Friendship x *G. tristis*. It takes 80-120 days to maturity with a flower spike length of 60 cm and floret colour Primrose.
- **Mayur :** Open pollinated seedling of the cultivar Sylvia, late season flowering. Plants are 82 cm tall with Lilac purple florets, dark purple throat. Cultivar is suited for cut flower and garden displays.
- **Nazrana :** This is another hybrid of Black Jack x Friendship cross. Produces flowers in 57 days. Spikes are very strong, charming with a length of 104 cm. Florets are cardinal red in colour.
- **Sapna :** Hybrid developed from a cross between Green wood pecker x Friendship. It requires 54 days to maturity from planting. Length of the spike which it attains is 83 cm. Florets are Barium yellow, with Primrose yellow blotch and Mandarin yellow margins. It has 17 florets per spike.
- **Suchitra :** is a hybrid from Sylvia x Jo Wagener. Plants are 90 cm tall. Florets are Camellia rose with stripes of vermillion and dianthus purple blotch.

Time of Planting : Based on research trials, planting schedules worked out at IIHR Hessarghatta showed that best planting time considering the quality of the flower stalk and number, size of florets, planting of corms during June, October and November is the best. June plantings of Gladiolus take shorter time for flowering and plantings during October, November take longer durations of 90 days.

Planting and Spacing : Corms are planted 4-5 inch deep in well prepared soil for Gladiolus culture. Spacing of 20 x 10 cm is given for the best density per acre which has been considered best for the spike yields. A density of 60000 – 80000 corms per acre has been recommended to be the best.

Gladiolus

Cultivation : Flowers appear after 90-120 days. These come up in flower spikes with 15-20 florest in it. Spikes are cut as soon as the first floret opens. Leaves of the plant are let undamaged, which are source of nutrition to the new corm. With the turning of the leaves to yellow these are cut and the plant is let to remain in the soil. Corms are then lifted and dried in shade.

Manuring and Fertilization: Research results at Marathwada Agri. University, Parbhani, suggest that the positive application of Nitrogen was enhanced with Phosphorous. The application rate of NPK was 50 : 50 : 50 kg per hectare. Application of this combination had an allround effect on the length of spike, number of corms and cormlets and the quality of flowers.

Flower Spike Availability : Period of spike availability under Indian conditions is different as per locations. In the plains of India, Gladiolus spikes after planting are available from the December month and continues till April. In the case of High hills of North India in the states of Jammu and Kashmir, Himachal Pradesh and Uttaranchal, spikes are available from August-October.

Propagation : A number of cormlets develop around the main corm. These are cultivated separately and develop in flower bearing plants after 2-3 years. Corms and cormlets are stored in cool dry conditions to break the dormancy. Best way of storing

Gladiolus corms under Delhi conditions is to keep in dry sand in a dark cool dry place.

Corm Harvesting: For harvesting corms special cultural operations are adapted for good yield of corms. Irrigation of the field is stopped immediately after the last harvest of the spikes and this is continued till the field is dry. After 1-1½ month of dry status of the field, corms are inspected in the soil, for harvesting. Corms are then harvested, cleaned and treated with DERASOL 0.2% for 30 minutes duration. Corms are then kept for drying in shade for 1-2 weeks. These are then stored in cool place and a temperature of 4-6°C is provided for better shelf life and storage.

Diseases: Major diseases which are observed are wilt, collar rot and storage rot of corms.

TUBEROSE

Tuberose is one such flower which has both beauty and fragrance which characteristic is not contained in all other flowers. Tuberose is cultivated in the world all over and India is one amongst those countries which produces Tuberose for cut as well as essential oil production. Tuberose culture is undertaken for cut flowers, Bouquets, Garlands, Landscape designing and essential oil production. Essential oil sells in the International market @ Rs 90,000 per liter. Tuberose culture in India has been a practice with the growers because of its easy and low cost cultivation and without any noted damage by pests and diseases. Tuberose is cultivated in India in the states of West Bengal, Tamilnadu, Andhra Pradesh, Maharashtra, Karnataka, Gujarat, UP and Delhi. In the state of Karnataka, Bangalore with its very mild climate is highly suited to the year round culture of Tuberose. Tuberose flowers are of two types i.e. Single and Double. In the Single group the varieties are – Calcutta Single, Rajat Rekha, Swaran Rekha. The double variety is Calcutta Double Pearl.

Climate : Tuberose is cultivated successfully with an optimum range of temperature of 20-30°C. Temperatures below 10°C and above 40°C are the stress temperatures at which no cultivation can be undertaken and there is no yield of flower heads for which this is done. Temperature of 21°C is very good for the spike initiation. Flowering starts in the months of July in North India and this continues till the end of December. Tuberose enter into dormancy in the month of March and April. Vegetative growth is too much in long days. Spike growth and flowering is associated with long days. Other effects on the growth of Tuberose are also contributed by low light intensity, which increases number of leaves, where as high light intensity adds to the flower quality of the Tuberose.

Site Selection : For the cultivation of Tuberose as a commercial crop, site should be such which has full sunlight through out the day and does not create any stress for low light. Site should be well drained, having no sign of water stagnation in the field or indications of creating water logging status. Light soils of Loam or Sandy loam are quite good with a pH of 6.5-7.5.

Preparation of Land : The soil is given a deep ploughing and is opened for exposure to direct hot sunrays, so that microbes and other weed grasses get killed and the field is nearly free from these. After 15-20 days regular ploughing is given to the already watered field and the soil is harrowed and planked. FYM and NPK fertilizer as basal dose are mixed in the soil prior to sowing of tubers.

Sowing Time : Normal sowing period is April-June in the plains of India and March-May in other North Indian states and West Bengal areas. In Bangalore this is done throughout the year and at any time of the month.

Manures and Fertilizers: IIHR Hasarghata recommends application of 12 tons per acre FYM and NPK in the ratio of 80 : 80 : 80 kg per acre. Phosphorous and Potash are applied as full doses as basal dressing and Nitrogen is split into 3 doses to be applied at the sowing time and after 2 months of planting and the last with the start of flowering. Application of Nitrogen in the form of Ammonium sulphate has been adjudged to be the best in comparison to Nitrogen from Urea or Calcium Ammonium Nitrate.

Propagation : Tuberose is propagated both from the tubers and seeds. Seed propagation is done by the breeders to develop new varieties but not for flower culture. Seeds take 10-14 days for germination. After germination the plants so formed are transplanted in earthen pots. For flower cultivation propagation is done from tubers.

Spacing : Tubers are spaced at 8-12 inch distance and are sown at a depth of 3-4 inch in dug up furrows. Plants come out from the tubers in about 20-30 days.

Irrigation : The crop is a summer season crop and it requires a good amount of water. Frequency of application of irrigation is kept during summer days at 5-6 days and during winter months from October onwards the frequency is decreased and irrigation time is increased for 10-12 days period.

Harvesting : Harvesting is started when the Tuberose has attained the flower maturity stage, which is noted with the appearance of two opened florets on the spike. Harvesting is done in the early hours of the morning when it is cool. Harvested stems are put in buckets containing water. Collected stems are then shifted to the collection center and hydro-cooled. These are then graded and packed for sale auction in the market. Grading of the spikes should be done on the basis of the type of the flower single and double, length of the spike and florets on the rachis, whether fragrant or not. All these characteristics carry a price in the market. Flower stems of Tuberose be stored at 10°C for increased vase life of the flowers.

Yield : Cut flower yield is based on the density of flower plants in the field. Generally the yield is 3-4 ton or 80000-1,50,000 spikes per acre and with an yield of 4-6 ton of tubers per acre. With the completion of flower harvest fields are not irrigated and the crop is set to dry. When the leaves get dried, tubers are dug out and treated with 0.1% Bavistin for 30 minutes to save these from fungal attacks on the tubers.

DAHLIA

Dahlia a plant of Mexican origin is cultivated from tubers. It has a very wide range of types, colours, size and shapes of the flower. Dahlia is a versatile flower extensively used for decoration, exhibition, garden displays. Dahlias are though very attractive and decorative with their size but the eye catcher is the "GIANT" whose flower size measures over 25.5 cm across. Giant Flower of Dahlia in India is a speciality in the exhibitions. Dahlia flowers remain in full glory for 4 months from the time of coming to flowering. Flowers of Dahlia in India are for decoration and colour spread in the home gardens. Dahlias are suited both for the pot and garden culture. Dahlias were first introduced in India during 1857 by the Royal Agri-Horticultural Society of India-Calcutta. Dahlias are cultivated during summers in the hills and in winters in the plains and remains in its glory for nearly 3-4 months.

Dahlia flowers are extensively being used for exhibitions, garden displays, flower arrangements.

Forms of Dahlia : Forms of Dahlias as classified are decorative, Cactus and Pompon. There are other forms too like Peony and Collarette types which are not cultivated in India.

Flower Pot Dahlias : Giant Dahlias which produce very large size of blooms are the most popular. To produce the largest flower size of the bloom, only the crown bud (top bud) is allowed to develop into a single flower, while all other flower buds are removed or rubbed off. It takes nearly 60-70 days from transplanting Dahlias and to produce large blooms. Liquid manure application instead of inorganic manure solutions is recommended for application to get bigger blooms (Swami Vinayanada, 1984). Selection of cultivars for growing of Dahlias for cut flowers and for pots is the same for both cultures. Suitable cultivars which branch freely should be the motto for selection of a cultivar for cut flower culture.

Cultivation : Dahlias love rich well drained soil and sunshine and require protection from frost and winds.

Spacing : Dwarf Dahlias are spaced 1-2 ft from plant to plant. Larger exhibit type Dahlias are spaced 2½ - 3 ft from plant to plant. Well grown up plants may require staking.

Climate : Dahlias are cold loving plants but require a good exposure of Sun light for best crop culture.

Soil Preparation : Soil is prepared thoroughly to give it a highly pulverized status for good root growth and high moisture retention. Soil is worked with a deep tiller to a depth of 1 ft.

Manuring and Fertilization: Application of manures and fertilizers to Dahlias have been recommended in two ways i.e. Basal and Booster doses. In the booster application dose recommendation is for a fertilizer of the composition 14 : 28 : 14 or 14 : 36 : 14 NPK @ 5-8 gm booster feed initially and at the bud formation stage 7-10 gm/plant supplemented with a dosage of 5 gm Magnesium (Swami Vinaynanda, 1984).

Propagation : Dahlias are best propagated from tuber division. However these can also be propagated from seeds and cuttings. Plants from tubers and cuttings are true to type. But plants from seeds differ very much from the parent plant. Dahlias take 80-90 days for maturity of flowers. Cluster of tubers attached to the collar of the plant are separated at the time of harvest taking care that the eyes are not damaged. These tubers develop as a new growth every year. Shape of the tuber is from round to oval and oblong.

Harvesting and Storage of Bulbs : After the flowering is over in Dahlias, crop is given a rest in the field itself, irrigation is stopped forth with, so as bulbs get acclimatized for storage. Harvested bulbs are cleaned and dried under shade and then put in cool chambers for storage. Dahlias go under dormancy in the plains. Instead of raising Dahlia plant culture from bulbs, commercial Dahlia cut flower growers rely much on the cutting stocks for the next crop.

Stock (*Mathiola incana*)

Stock is an annual flower cultured for its cut flowers for flower arrangement and have very fascinating colours. Flower spikes which are 6-7 inch long are full of 1 inch florets arranged in clusters. Flowers are fragrant.

Colours : Red, Violet, Purple, White, Blue.

Cultural Notes : Pick the spikes when 25-50% of the florets are open, but before flowers at base begin to fade.

Time of Planting : It is a cool season flower. In India it is planted during the winter season in the plains and from March – April in the high hill zones.

Propagation : Stock is propagated by seeds.

Post Harvest Treatment : Before arranging, re-cut stems, remove foliage that will be under water and put in a mixture of warm water and floral preservative for a few hours or over night. Place the flower arrangement, away from direct sunlight, heat vents, air conditioners to increase the vase life. Water and remove dying blooms and foliage daily. To increase vase life, re-cut the stems every 4-5 days and clean container thoroughly. Re-arrange flowers adding a mixture of warm water and floral preservative.

FREESIAS

Freesias originated in the Cape of South Africa. Freesias are mainly being cultured as cut flowers from corms. Cut flowers of trade are 25-35 cm long. Freesias are sold as potted plants. Critical factor in Freesia culture is temperature. Day temperature 16-17°C and night temperature 10-16°C are maintained. Freesias take 14-17 weeks for cut flowers to mature and 9-11 weeks for potted plants. Potted Freesias are put in pots which are 4-8 inch deep putting 6-15 corms.

Planting Media : Corms are either planted in ground beds or raised benches which are atleast 25 cm deep. Corms are planted 5 cm deep with a plant density of 97-120 corms per sq. meter.

Cultivars / Varieties : There are single and double flower cultivars of Freesias. Regular Pot plant cultivars of Frees-ias for potting are Amadeaus, Blue Heaven, Flandria, Kayake, Marianne, Purple Rani, Sarnia, Vulcano, Yvonne. Other cultivars are Gombey, Popeye, Smarty and Suzy.

Forcing for Flowering : Forcing Freesias for early flowering of potted plants is to produce marketable plant in 60-80 days from planting with an average plant height of 25-41 cm. The goals of shorter duration with dwarfness in potted plants is not possible. However this has been achieved with newly released cultivars from Netherlands known as "**Easy Pots**". These four cultivars do not require plant growth regulators to reduce the height. On arrival corms of Easy cultivars are planted immediately.

Other Cultivars : Which are received, are subjected to pre-cooling at 13°C for 45-49 days with a high RH, good air circulation but no ventilation. Corms are not to be stored for more than 49 days. This can cause corms to pupate i.e. they will form a new corm instead of a shoot. Corms pre-cooled at 13°C are planted immediately for cut flowers. For potted Freesias, height of the plant is restricted by giving 1 hour dip in PACLOBUTRAZOL at a given concentration before planting, different for each cultivar.

Planting : Corms after pre-cooling or dip in Paclobutrazol are planted immediately 2.5 cm deep. After planting keep the medium moist but not wet. When plants start growing use 14 : 14 : 14 or 20 : 20 : 20 fertilizer. Freesias are very sensitive to fluoride which induces Leaf scorch. Because of this do not use superphosphate or other fluoride containing amendments in the planting medium or water.

Post Harvest : Cut flowers immediately when the lower most floret opens up. Little or no storage is advised. For very short storage hold at 0-2°C and 95% RH levels.

Light Intensities : Under the covered Freesia culture the light intensity of 27-54 kLux of medium and high intensity is required in the Green House.

ORCHIDS

In the Monocotyledonae Orchids with 600-800 genera and 25000-35000 species (Arditti, 1981) represent the most highly evolved family with 40% of the monocotelednous taxa. Most of the Orchids are perennials with simple leaves and are distributed in all continents but for Antarctica majority of the Orchids are epiphytes or terrestrial and a few saprophytes. These occur in humid tropical forests of South and Central America, Mexico, India, Srilanka, Thailand, Myanmar, South China, Malaysia, Philippines and Australia, New Guinae. Modern Orchid trade and industry is there for Brazilian Cattleyas, Dendrobiums, Cymbidiums, Vandas and Phaleonopsis as cut flowers stems or potted plants. Orchid culture in India is restricted for export because of Indian Governments ratification of the recommendations of Convention of International Trade in Endangered Species of wild fauna and flora and placing Orchids in Appendix–II which puts an embargo not on its culture but exports through permits. India has two conservation and Germplasm centers at Shillong in Meghalaya and Yercud in South. States of Assam, Meghalaya, Arunachal Pradesh, West Bengal, Karnataka and Kerala, Sikkim are most suited to Orchid culture. Orchid culture in India is not organized and deserves high attention from the flower growers and the hobbysts so as to learn and then compete in the trade. Temeperate Orchids are terrestrial and the Tropical one's are epiphytes. Orchids have the habit to grow under diffused light system thus require very less light under the controlled conditions in the green House and are provided with shade nets for that.

Orchids are well known for their curative and aphrodisiac properties too and have been appreciated for their value in Sanskrit and English literature because of medicinal values and therapeutic properties. Many species of Orchids are used in folk medicine in India. Theophrastus (372-287 BC) and Linnaeus were interested in Orchids principally as medicinal plants. Orchids cure a variety of ailments viz. muscular pains, headache, stomach ache, bone fracture, general debility and weakness. These are also used to stimulate hair growth, preventing baldness, as skin rejuvenator. Important medicinal Orchids are :—

- ***Phaius tankervilliae*** : A terrestrial specie, the rapeutically important Orchid.

- ***Saccolabium papillosum* Lindl. :** Distributed along the Himalayas from Kumaon eastwards to Arunachal Pradesh. Roots of this are used as a substitute for Sarsaprilla.
- ***Bulbophyllum careyanum* (Hook) :** Species is distributed in the base of Himalayan ranges tropical climates from Sikkim eastwards to Arunachal Pradesh.
- ***Malaxis acuminata* D. Don** : Dwells in Pine forests and distributed from Himachal Pradesh to Arunachal Pradesh. Dried bulbs of this are used in the preparation of Chyavanprash and also to cure Tuberculosis.
- ***Dactylorhiza hatagirea* D. Don :** Species is native of Spiti valley. Boiled extract of the tubers mixed with milk is given for treating bone fractures. Fresh extract is recommended for curing cough and stomach ache.
- ***Vanilla wightiana* Lindl.:** is distributed in the southern Seshachalam hills and is used for curing of cold, cough, skin burns, ulcers, viral fevers and bone fractures.

Orchids here in India are not being exploited for cut flower export, which has the biggest potential as a commercial floriculture as cut flower in the export markets of Europe and Japan. The other utility is in the development of trade for medicinal properties of Orchids which stand very much explained in the scripts.

Classification : Orchids can be classified as Monopodial and Sympodial depending on the habit of their growth.

- **Monopodials :** have a main stem which continues to grow year after year producing inflorescence from the leaf axil. Phaleonopsis, Renanthera and Vanda can be placed in this category.
- **Sympodial :** have a main stem, which terminates growth at the end of each season. New shoot which grows from the base forming its own bulbous stem which stores the plant food, is called pseudo-bulb, which ultimately flowers. Catleya, Dendrobium, Cymbidium form this category.

Orchids can further be classified as Ground and Epiphytic, suited to both the tropical and temperate regions. Indian species of Orchids in both the groups as per their characteristics are :—

GROUND	EPIPHYTIC
Arundina	Aerides

Calanthe	Bulbophyllum
Cymbidium	Cattleya
Paphiopedilum	Coelogyne
Phaius	Cymbidium
Pleione	Dendrobium
	Odontoglossum
	Oncidium
Rynchostylis	
	Thunia
	Vanda

Further classification based on the geographical culture and availability of the species in the tropical and temperate regions is more clear, detailed as under —

TROPICAL	TEMPERATE
Arundina	Aerides
Bulbophyllum	Bulbophyllum
Cattleya	Calanthe
Cymbidium	Cymbidium
Dendrobium	Dendrobium
Oncidium	Paphiopedilum
Rynchostylis	Phaius
Vanda	Pleione
	Odontoglossum
	Thunia
	Vanda

Orchid Culture : Orchids in nature grow away from the sunlight but in diffused light of the sun under the shade of the tree's. To provide similar growing conditions of nature, under artificial culture in the field, Orchid houses can be developed with the local know-how and availability of materials like Bamboos and other market materials like glass, fiber sheets. A note of caution be understood that Orchid culture of the tropics is different from that of the temperate region. Orchids of the tropics enjoy the warm temperature, humidity and those of the temperate region cold temperature with low humid conditions. Best temperature range for Orchid culture is between 18-30°C, with proper ventilation in the orchid chambers. For Tropical Orchid culture, the Orchid chambers are provided in the center with a tank like structure which maintains the humid conditions.

Manuring and Fertilization : Orchids in nature obtain their food from the natures hosts like trees which provide every type of nutrient feed both macro and micro from their bark and also from the air. Inorganic micro-nutrients like calcium, magnesium, iron, manganese, boron, copper, zinc and macro nutrients like Nitrogen, potassium are obtained from the tree bark on which these grow. For commercial controlled culture now a days very good granular and soluble elemental mixtures are available in the market. Orchids are very slow growing plants as such plant nutrient mixtures be selected such, which contain slow release mixtures of nutrients like Osmocote. Recommendations are for the application of Nitrogen in good quantities during the vegetative growth period and then the application of Phosphorous and Potassium.

Propagation : Cuttings, Lateral buds and the roots produced from the tubers are the practices being adopted for the propagation of Orchids. However, Vandas can be even multiplied by air-layering (Gootie), when a cut is given 20-25 cm below the apex and moist sphagnum moss ball is put around it and tied. Seed germination is not possible because of too small a seed size. The best method these days is the production of seedlings plants through Tissue culture which alone is meeting the huge demands.

Insect Pests : Most common insect pests of the orchids are thrips, aphids, spidermite, mealybugs, weevils, snails and slugs. Harm created to the orchids is in the form that these feed on younger tender shoots, suck the sap and damage the buds and become carriers of diseases.

Diseases : Fungal and Bacterial diseases which have been noticed are Leaf spot, Leaf blight, Collar blotch, Collar rot, Orchid wilt.

WORLD TRADE

Trade of Orchid cut flowers and potted plants is ever growing. During the year 2000 a trade of 128 million USD in cut flowers and 23 million USD in potted Orchid plants has been reported according to United Nations Comstats (Floraculture 2002, October). Thailand in the Asian region with a trade of 50 million and 4 million US$ in cut flowers and potted Orchids respectively dominated. Thailand has 2240 Ha in Orchid culture. Return per hectare comes to 24110 US$, which is one of the best for any crop. Other Asian players in the Orchid trade are Singapore 7.7 million USD, Malaysia 2.8 million US$. Italy in Europe comes next with a trade of 6,52,000 US$. Importers of Orchids are Japan, Italy, France and Germany. China largest producer of potted Orchid plants of Cymbidium, has only domestic market for its produce. Largest exporter of Orchids in the world is Taiwan. World rankings in export trade of Orchids stand are Taiwan, Thailand, United Kingdom, Italy, Japan, Newzealand, Brazil, India, Malaysia, Ivory Coast, Srilanka, Peru, Mauritius, Vietnam, Singapore, Phillipines, Madagascar, China. Most valuable blooming potted plant Orchid in the Netherlands auctions is Phaleonopsis.

Developments in Orchid Trade : Development of Orchid trade the world over has been necessitated by the consumers with their preference for its size, quality, colours and the vase life. However all round development has lead to the following in the trade:—

- Growing demand of Orchids has lead to the increased size of the Farms particularly in Thailand, Malaysia, Costa Rica, Panama where the average size of the Farms is 8 Ha.
- Greater part of investments is directed to Orchid culture.
- Breeders have directed their research in Dendrobiums to the naming of un-named varieties, plant improvement through crossing for long stems, larger flowers and a wide range of colours.
- Increasing vase life. Many cut Orchids have a vase life of one month.

World Orchid Trade : In the world trade of Orchid Imports as cut flowers and potted plants Japan is the biggest importer followed by Italy and France, Germany, Britain, Netherlands, Austria. World trade has grown from 150 million US$ during the year to 230 million US$ during the year 2001 (Floraculture International, Oct. 2002). China is the largest producer of potted Orchids (Cymbidiums) but its produce is locally consumed and there is very little for export. Regarding Orchid export trade Taiwan is the largest orchid plant exporter. Other Orchid plant exporters in the world are Thailand, United Kingdom, Italy,

Japan, Newzealand, Brazil, India, Malaysia, Ivory Coast, Srilanka, Peru, Mauritius, Vietnam, Singapore, Phillipines. In Europe the domestic production of Cymbidium and Phaleonopsis is mostly for domestic markets. In the market of Netherlands the most popular and demanding potted blooming plant is Phaleonopsis. With Japan as the largest import market for Orchids, USA stands the other as the largest importer of Orchids in the world.

Developments in Orchid Culture and Trade : Culture of Orchids has expanded so much in the recent past that the latest development to observe are many faceted, giving an impression that this particular crop commands value in the market and an interest in the growing countries with the farmers. Various developments which have been observed for the creation of this impact are —

1. The increasing size of Orchid Farms. In Thailand several Dendrobium Farms are over 30 acres. Malaysia, Costa rica and Panama have Dendrobium Farms above 8 Hectares or 20 acres.
2. The second development in this is the investment of large amounts as commercial projects.
3. Breeding of new Hybrids in Dendrobium species, with a straight emphasis on the development of new cultivars with long stems, larger flowers and a wide range of colours.
4. Orchid plantations at high altitudes are producing cut Dendrobiums and Oncidiums with an extended vase life, where many cut orchids have a vase life of one month.
5. Orchids are the flowers and potted plants which are to be kept at separate temperature regimes and away from other plants producing rich amounts of Ethylene. Thus now the distribution system has been designed in a separate fashion where these are transported and marketed separately.

LILIES

Depth of planting in lilies is the critical cultural factor. Some of the lilies root from the base of the bulb while others from the stem but above the bulb. As such stem rooting lilies are planted deeper than the bulb rooting one's. Thus a depth of 6-8 cm for the bulb rooting and 12-15 cm for the stem rooting is quite good. Best planting time is after the dormancy of the bulb is over after nearly 2 months from harvest. Lilies under hill condition are planted during February-March and during September under plains or Delhi conditions.

COMMERCIAL CUT FLOWER LILIES

Common Name	**Bot. Name**
Japanes Easter Lily	*Lilium longiflorum*
Tiger Lily	*Lilium tigrinum*
Belladona Lily	*Amaryllis*
Torch Lily	*Kniphofia*
Blue African Lily	*Agapanthus*
Calla Lily	*Zantedeschia rehmannii*
Peruvian Lily	*Alstroemeria*
Tulips	Tulip spp.
Freesia	
Hemerocallis	

Lilies of Commerce : In the trade of Lilies for cut flower and potted plants, trade is developing in the same way as for orchids. Lilies has been described as commercial floriculture plants with the research aimed at increasing the array of Lilies that could be adapted to pot culture. World wide Lilies are mainly grown as cut flower crop. Hybrid Lilies are a diverse group of plants that can be forced for year round sales as cut flowers but choice for pot plants is also increasing, with a very wide choice of improvising cut flower Lilies for pot culture too. This is being envisaged because of relative ease of Lily production and high crop value per square meter and increasing variety selection and better height control.

Hybrid Kinds of Lilies : Lilies have been placed in three distinct categories of Oriental Hybrids, Asiatic Hybrids and LA's or Longiflorum / Asiatic Hybrids.

- **Oriental Hybrids :** are known for their fragrance, large lush flowers maqinly in pink, white and red and many other combinations. Main varieties are - Alliance, Aubade, Helvetia, Muscadet, Star Gazer, Tom Pouce. Pot Oriental Hybrids in this are Star Gazer, Monalisa. Crop cycle is for 90-110 days.
- **Asiatic Hybrids :** have little or no fragrance and are with smaller leaves and come in any colour except blue

and black. Amarone, Colosseo, Gironde, Tresor, Vermeer, Vivaldi are Asiatic group varieties. These have a much shorter cycle of 50 days.

- **LA Hybrids :** LA hybrids were first introduced in the world market in 1987. These have become highly popular in recent years. LA hybrid flowers have a good bud count, feature lateral flowers and can reach lengths of 80 cm. Second generation LA hybrids can be characterized as more Asiatic than Longiflorum. Alladin Dazzle and Salmon classic are examples of this. Third generation LA hybrids are characterized by huge flowers, almost as big as Orientals. These Lilies can grow up to 130 cm. LA hybrids require few days in Green House than other Lilies and these are less susceptible to low light conditions. They are more resistant to Botrytis, Fusarium and Pythium, buds open rapidly. Longiflorums are Interspecific Hybrids of the Easter Lily and Asiatic Hybrids cultivars. These combine good characteristics of each of the groups and have excellent foliage, large prominent buds and a medium length crop cycle with colours of mainly pastel pink, red and yellow. Brighter colours are also available. Bestseller, Ceb Dazzle, Fangio, Royal Dream, Royal Fantasy, Royal Parade, Royal Perfume, Salmon Classic, Samor are the cultivars of this group.

Planting Time : Lilies have two seasons of planting i.e. summer / rainy season and winter or cool season. Planting seasons are based on the availability of cut flowers. Flowers which are in the market during February to April are the spring flowering cultivars and are sown during the months September to October. Lily flowers harvested during May-September are sown in the field during February to march.

Bulb Forcing : Purchase of Lily bulbs be ascertained whether these have been forced or require forcing treatment prior to planting. If the bulbs have been forced and arrive in freezing condition these should be thawed first in a cooler. Freshly harvested Asiatic hybrid bulbs are forced by pre-cooling for at least 6-8 weeks at 1-2°C. Orientals and LA's are cooled for 8-10 weeks at 1-1.6°C. Additional time at these temperatures can be used for short term holding.

Propagation: All the Lily species are propagated through bulbs but for Tiger Lily which is propagated through scales and seeds.

Spacing and Density : All the Lilies are spaced at 1 x 1 ft, thereby accommodating nearly 42000-45000 plants per acre.

Pests and Diseases : Nematodes and Thrips are the pests which damage thc Lily crop. Rhizome / Bulb rot, leaf spot is the disease which is most serious in Lily culture.

Post Harvest Lily Quality : To maintain post harvest quality of vase life for Lilies and their freshness for auction sales, latest usage of Giberraline on the plant has been studied and recommendations made for maintaining the post harvest quality of Lily flowers. Giberalline product GA 4+7 available under the properietory chemical FASCINATION is a powerful inhibitor of leaf yellowing. It also improves flower longevity. It also allows hybrid Lilies to tolerate short term dark cold storage (2-4°C for up to 2 weeks). Foliar spray of 25 mg/liter of the Giberalline 4+7 component within 2 weeks of harvest is sufficient (Floraculture – May, 2003).

Growth Regulation for Potted Liliums : Market research has recommended usage of growth restriction for pot Lilies for best auction sales or returns by the farmer. This is best achieved by controlling the height through dipping treatment of the bulbs. Growth regulators are Paclobutrazole and Unicon-azole available under the brand names BONZI and SUMAGIC in the market. Dipping time is for one minute and the concentration is as under:–

	GROWTH REGULATORS	
TYPE OF HYBRID	BONZI	SUMAGIC
Oriental Hybrids	200 – 300 ppm	5 -10 ppm
LA- Hybrids	150 – 250 ppm	2.5- 5 ppm
Asiatic Hybrids	50 – 100 ppm	1- 2.5 ppm

CALLA LILY

Calla Lilies are an up-coming cut flower and potted plant crop world wide. The characteristic has arisen because of its elegance and class in flowers, which has brought fame to this unknown flower. Uniform culture of Calla Lilies require special attention to media selection, water management, forcing temper-atures as well as disease control management. Calla Lily is :—

Zantedeschia rehmannii	Calla Lily
Zantedeschia elliotiana	Calla Lily

Propagation : Zantedeschia lilies are propagated through seeds and tubers. Tissue culture is also gaining popularity in producing plantlets in Netherlands and Newzealand. Calla growers however go for the purchase of tubers from specialized growers, who take every care in providing the best of the treated tubers. Calla Lilies sold in USA are 2 year old true seed F1 hybrids. Treated seed nearly eliminates viral, fungal, bacterial pathogens, thereby greatly reducing disease. Seed grown Calla Lily tubers has many eyes thus these plants will have high number of flowers and leaf sprouts. Zantedeschia tubers are available in 5 grade sizes of 12-14, 14-16, 16-18, 18-20 and 20 + cm circumference.

Crop Cycle : Calla Lily has a crop cycle of 9-13 weeks for complete production of cut flowers and potted plants.

Temperature Management : Temperature management is the critical part of management in this culture to bring the crop at the right time and also to control the plant height. Calla prefers a day temperature of 17-29°C and a night temperature of 12-17°C.

Calla Lily Culture : Success in Zantedeschia culture for potted plants and cut flowers is dependent on critical disease management aspects. Seed/Tuber suppliers supply planting material duly treated with GA and fungicide-bactericide solutions for disease control. Calla's respond to GA concentrates is up-to a strength of 500 ppm, but treatment of 100-150 ppm concentration has been considered best with GA3 (ProGib) or GA 4+7 (Promalin) solutions. Potted plant culture of Calla's can utilize tubers of various sizes depending on the size of the pot. Zantedeschia farmers prefer container plantings for this crop, because of the ease of moving containers. Open air culture of Zantedeschia suggests change after every 3 years. In pot culture tubers can be planted singly or in multiples as under —

Pot size	Tuber size	No. of Tubers Per Pot
11 cm	4.5 – 5 cm	1
15 cm	18 – 20 cm	1
15 cm	4.5 – 5 cm	2
15 cm	4 – 4.5 cm	3

Tuber Storage : Un-pack untreated tubers, expose for ventilation at 15-18°C to dry off abrasions, cuts. Inspect stock and discard soft, rotted tubers. For storage more than a month, hold tubers at 7°C at 80% RH level, with a good circulation of air, which is highly necessary for Calla's.

Disease Control : Disease control management is the most vital aspect in Zantedeschia flower culture. Main Bacterial diseases affecting are *Erwinia carotovora, Penicillium*, Fungal pathogens attacking are *Pythium* and Rhizoctonia. Viral infections are due to Dasheen Mosaic virus, Konjac Mosaic virus and Zantedeschia Mosaic virus found in Asia.

- ***Erwinia carotovora*** : Leaves and stem turn dark green and exhibit wet foul smelling rot and the plant falls. Bacteria are present in soil and plant parts, often also found in ground water. Antibiotics and other disinfectants can kill bacteria but not destroy the pathogen. Hot water treatment kills most of bacteria in tuber, but once it is planted, is again surrounded by bacteria. Prevent plant and tuber damage from wind and herbicides. Avoid Pythium and Rhiozoctonia infections, which act as gateways for Erwinia. Use moderate rates of Nitrogen.
- ***Pythium*** : causes root rot, interferes with plants water uptake. Control is difficult. Proper irrigation and drainage are important components of preventing an infection.
- ***Rhizoctonia*** : affects the shoot at its emergence point. Fissures are visible on the underside of tuber. Preventive soil treatment is possible.
- ***Pennicillium*** : infects tubers during storage. Most common causes of this disease are damaged tubers, high humidity levels in storage chambers. Insufficient air circulation is also one of the causes of this disease. Ensure proper air circulation and disinfection of tubers.
- ***Botrytis*** : affects flowers and foliage. Spraying frequently before flowering is sufficient to prevent damage.
- ***Viral Infections*** : Viral infection symptoms are spotted, mis-shapen leaves and flowers. Infection is transmitted by Aphids and Thrips. Weekly sprays of crop near flowering time with mineral oil supplemented with insecticide keeps the crop free from viral infections and attacks. Keep crop weed free.

Post Harvest Management of Lilies : Oriental and LA hybrids have the strongest root system in the Lilies than the Asiatics, which are very week. Harvest Lilium stems by cutting and not pulling. Harvest Lilium stems in the morning hours and at full maturity when there are 10 + flower buds when there are at least three coloured buds; and stems with 5-10 buds, there are at least two coloured buds and stems with buds less than five buds at least one coloured bud is there. Harvesting in the morning is done to prevent drying off during the day. After harvest cool stems to 1-2°C as quickly as possible. Once cooled down, stems are then bunched. During bunching remove leaves from the lower 10 cm length of stem. Store bunched stems upright in clean water in the cooler to re-hydrate them. Asiatic hybrids may be treated with STS (Silver Thio-sulphate) preservative. However STS will damage other Lilies.

ALSTROEMERIA

Alstroemeria develops from the vertical shoots of an underground rhizome. lateral rhizomes can also be generated by the main rhizome, these also produce shoots. Major part of Alstroemeria development happens underground, so soil temperature is an integral part of the growing process.

Soil : Alstroemeria grows in all types of soils but like well aerated and drained soils, high phosphate levels, which is very important for its consistent growth. Sandy and clay loam soils are best for Alstroemeria culture. In soils with pH above 7.0, which are generally defficient in iron or manganese, Alstroemeria plants indicate this, through the yellowing of the leaves. Supplement iron deficiency with the application of 6-8 gm EDDHA per sq. mtr area.

Spacing : Alstroemeria is planted in 1.02-1.19 mtr wide beds giving a plant to plant distance of 34-40 cm depending on the variety. Rows are spaced 40-50 cm apart depending on how quickly the rhizome grows towards the edge of the bed.

Temperature : Maintain a temperature of 13°C during night and 14-16°C during the day after 6 weeks of planting. During summers temperatures between 17-20°C are best for Alstroemeria. Temperatures 10-14°C are optimal during winters depending on the cultivar. Temperatures below 10°C, means slow plant growth.

Alstroemeria

Humidity : In general humidity levels recommended are 70-80%. Higher RH levels of 90% may lead to taller stems with weaker leaves.

Post Harvest Treatments : Alstroemeria is harvested by pulling the stems. After grading place the flower stems in a solution with Gibberellin to prevent leaf yellowing.

CHRYSANTHEMUM

Chrysanthemum is the most popular flower after Roses for cut flower, bedding, pot plant and flower arrangement. Spread of Chrysanthemum has been recorded from China to other parts of the world. It was mentioned by Confucious as early as 500 BC. From China it reached Japan via Korea more than a thousand years back. In Japan it went under a great transformation in the production of very many types giving a birth to regular and irregular blooming types and was declared as national flower of Japan. In the UK it is regarded as one of the best flower exhibits. In the United States it is one of the best cut flower and a Dollar earning flower. Blooms of Chrysanthemum last for more than a month. Culture is comparatively easier than Roses as this has no typical soil or light requirements. Chrysanthemum is a short day plant but cultivars which are day neutral also form a part of the varieties. Wide range of colours sizes, shapes are present in this group. Tiny flowered or button hole size flowers are called Pompons.

Varieties : Varieties of Chrysanthemum are determined by the types, which have been classified as Large flowered, small flowered. The type of flowers classified in the large flowered Chrysanthemum are *"Incurved, Incurving, Reflexed, intermediate, Irregular, Quilled, Spoon, Anemone, Single, Semidouble"*. In the small flowered group classification is for *"Button, Korean Single, Korean Double, Decorative, Pompon, Stellate, Semi-quill"*.

Culture : Chrysanthemum culture responds differently to the environment both underground and aerial. The plant has a fibrous root system, which is highly sensitive to water logging, pest and disease attacks, but fairly tolerant to drought. Growing / bedding/ potting media should be such should take care of aeration, water retention and also supply of plant nutrients to the plant. In the aerial environment of Chrysanthemum requirement light, temperature and humidity are the chief factors which are considered for good culture. Long days, higher light intensities, high temperatures lead to more vegetative growth of the plant. Critical point of night length is 9½ hours. No sooner this limit is crossed, flower bud initiation starts. Growth is optimum at 15°C and the plant remains stunted when temperatures cross 25°C. Range of temperature between 16-18°C has been found to be the best for this flower. Based on the temperature requirement of Chrysanthemum, Cathey has classified into 3 distinct groups, which are :—

- **Thermozero :** Flowering at any temperature between 10-27°C but most consistently at 17°C night temperature.
- **Thermopositive:** Continuous low temperature between 10 and 13°C inhibited or delayed bud initiation. At higher temperatures 27°C bud initiation is rapid but flowering is delayed.
- **Thermonegative :** Bud initiation occurs in low to high temperatures between 10-27°C, but continuous high temperatures delay bud development.

Soil : Rich, deep, well drained soils are best suited for Chrysanthemum culture. Loam soils are the best under Indain conditions which have an equal quantity of clay and sand in it well conform to the required qualities. Chrysanthemum thrive in good sunshine under open culture.

Varieties : Chrysanthemum varieties are different for cut flowers, potting plants. Important Indian Chrysanthemums in both the categories are :

A. CUT FLOWERS

- **White :** Baggi, Moud Jappiers, Himani, Gajra, White Quills.
- **Yellow :** Nanako, Aprajita, Ajay, Yellow prolific, Sonali Tara, Basanti, Mountaineer.
- **Purple :** Shyama, Winsom, President Vizar, Harvest, Home Flirtation, Nilima.
- **Bronze / Red :** Flirt, Ratna, Jaya, Jayanti, Molly.

B. POT VARIETIES

- **White :** Himani, Gajra, Fishtail, Mrs. C. Tolly,
- **Yellow :** Kundan, Aprajita, Mayur.
- **Bronze / Red :** Bindiya, Ratna, Rita

Spacing : Spacing depends on the varieties which is different for the Exhibition and pompons. Spacing in the case of Pompons is 30-45 cm from plant to plant while in the case of exhibition type or large flowered species it is 45-90 cm. Stakes are provided to the plants in the soil and in the pots to bear the plant and flower, because stems of Chrysanthemum are very fragile and weak. Pompons do not require any staking.

Stopping : This is one special cultural practice, where in the formation of break bud is prevented for Chrysanthemum flower culture. This is started when the plant is 6-8 inch in height. Side shoots are allowed to grow until these develop buds. Number of such side shoots is restricted to 8 or so. In case of large exhibition blooms, the number may be restricted only to one or at the most 2-3 shoots. All the other shoots and the axil growth on the selected shoots are rubbed off as these appear.

Propagation : Chrysanthemum plants do very well when propagated from cuttings. Cuttings reduce the maturity period and aid in bringing the produce at the earliest in the market. Division from suckers is also well practiced, but takes longer duration for maturity than rooted cuttings. Under Indian conditions Chrysanthemum is susceptible to decay during hot and moist weather of the monsoons. Plants thus be saved from this and protected by keeping under shade. February to April are the best periods for the propagation of suckers and July – August are the best months under Indian conditions for propagation of cuttings.

Manuring and Fertilization : Chrysanthemums respond too much to good nutrition levels both for cut flowers and also for potted plant culture. Recommended NPK levels at the YS Parmar University of Horticulture and Forestry, Solan are 160 : 80 : 80 kg application per acre with a basal dose of 20 ton per acre of well rotten FYM.

Cut Flower Numbers : To increase the number of cut flowers per plant, the plants are subjected to another cultural operation of pinching, which gives Chrysanthemum plant a bushy appearance and creates conditions for maximum number of cut flower branches per plant. The process is taken up when the plants are 6-8 inch tall and this is immediately after the cuttings have taken a start during the months of August and September. The second and third operations are done after 30-35 days of the first, keeping a note that only one well developed flowering bud is kept per branch and others are disbudded.

Harvesting : Harvesting of spray, pompons, exhibit type cut flowers is done very differently for each of the types. For spray chrysanthemums when the lower 2-3 flowers have opened up, it is the best time to harvest. In standards, it is the un-opened stage when harvesting is done. In standards when the outer ray florets start unfurling, it is the best time. Harvesting is undertaken in the cool hours of the evening or morning and also keeping in view the market of the area, how far it is. For distant markets flowers of both spray and standards are harvested at half open stage and immediately put in keeping 200 ppm solution of 8-HQ Citrate + 2.5% Sucrose for 24 hours at 21°C.

Flower Yield : Each plant gives a minimum of 4 blooms in its planting cycle and expected yield of blooms is around 2.10 – 2.30 lakh blooms per acre.

Pests and Diseases: Common pests of Chrysanthemum is Aphids, beetles and bugs. In the diseases powdery mildew and wilt are the other diseases which are of concern.

PANSY

Pansy is an annual in the hotter climate zones but a perennial flower in the temperate cool zones. Pansy is popular because of its very wide range of colours, flower size and its usage as cut flower and a potted plant, a plant for landscape designing and all the more as a presentation in Bouquets forms. Breeding of Pansies have been very directive for the production of cultivars having different colours, flower size, number of flowers per plant, temperature tolerance to the respective zones of culture, added new dimensions to commercial culture of Pansies. Under Indian conditions Pansy is a cool season crop and is cultivated during the winter months.

History of Pansy : Pansy is Viola tricolor and is the progenitor of modern Pansy for cut flower, landscape designing and flower arrangement. With the development of modern pansy (*Viola wittrockiana*) flower has added colour shades of pink, rose, orange have gained high popularity among the society of flower lovers. Breeding of Pansies as of today is more concentrated in USA, Germany, Japan and is directed on the aspects of quality – such as vigor, free flowering and heat tolerance. Pansy flowers both of single and double types have red, purple, blue, bronze, pink, black, yellow, white, lavender, orange, apricot and mahogany colours. Pansy is the only unique flower with all the desired colour ranges whether in pure singles called "CLEAR" and multi-colour of flowers. Multicoloured flowers that have a very dark blue/black centers are known as "Blotched" or "Faced". Based on flower sizes there are three groups of Pansy varieties —

- **Large :** 3 ½ - 4 inch dia. bloom.
- **Medium :** 2 ½ - 3 ½ inch dia blooms
- **Multiflora :** 1 ½ - 2 ½ inch dia blooms

Seed Germination Requirements : Pansy has a typical and peculiar characteristic of having two types of seeds not contained in any other plant. Pansy has Traditional and Primed seeds. Traditional seeds are not treated before sowing to enhance germination. Primed seeds are physiologically treated to start the germination process. Advantages of Primed seeds are —

- Germination percentage is higher.
- Germination of the seeds is very fast.

- Seeds are less sensitive to higher temperatures during germination.

Substrate for Pansy seed germination should have adequate aeration and good moisture retention. Pansy seedlings are sensitive to higher concentrations of soluble salts. High Phosphorous levels cause seedling stretch, thus Phosphorous levels in substrate should be low. Use of fertilizers low in Nitrogen and Phosphorous to avoid stretch is recommended. Fertilizer formulations with Nitrogen derived from Potassium Nitrate, Calcium Nitrate, Magnesium Nitrate are recommended, as these supply micro-nutrients like Ca, Mg, pH of substrate should range between 5.4-5.8. Optimum temperature for Pansy seed germination is 20°C. Cool temperature range of 15.5-21.0°C is best for seed ling growth. Seed takes nearly 14 days for germination or the first leaf to come up after the emergence of radicle and cotyledons. Seedlings are ready after 45 days from sowing.

Growing Climate : Pansy plant attains a maximum height of 6 inch and grows best in shade.

Planting : Nursery raised Pansy seedlings are trans-planted at a spacing of 9 x 9 inch both from plant to plant and row to row. Pansy plants are planted with a near density of 75000-80000 plants per acre for cut flowers culture.

Propagation : Pansy is propagated by seeds either making a direct sowing in the seed beds or putting these into plugs. Plug raised seedlings are commercially very much in demand in the western countries. In India this cultural procedure is yet to take time.

MARIGOLD / GAINDA

Tagetes is a very popular flower, easy to grow and as a garden edging. African Marigolds are tall in comparison to French which are dwarf. It takes a cycle of 8-12 weeks for full flowering. Marigold has been described being the native of America and Mexico. Marigold under Indian culture and conditions is cultivated only for offering to deities and Goddesses of the temple and home. It is used in the form of loose petals and in Garlands. Presently it is being used for the preparation of natural colours and some of the organizations are doing the commercial job of making natural colours of red, yellow and orange from Marigold. During the months from July to October demand of Marigold is heavy in the market and rates per kg have even touched Rs 15-20 per kg. Apart from this Essential oil of Marigold and extracts are also used medicinally. In the North Indian plains Marigold is cultivated through out the year and plantings are undertaken during winters, summers and rainy season. Cultivation of marigold is confined to the adjacent areas of the metros of India, where it should not take more than 3-4 hours for transport of harvested produce to reach the market. Marigold belongs to genus Tagetes and various cultivars in this are :—

Tagetes erecta	African marigold
Tagetes petula	French marigold
Tagetes minuta	Wild marigold
Tagetes signata pumila	

Varieties

- **African :** Giant Double, African Orange, Climax, Local
- **French :** Rusty Red, Butter Scotch, Red Brocade

Colours : Available colours in Marigold are yellow, Orange, Lemon, Dark Brown and also combination of these colours. White Marigold have been bred and developed by **Burpee's of USA,** which provides a very healthy differentiating look in the beds and are the suppliers of this seed variety.

Propagation : Marigolds are propagated through seeds. Seeds are raised in a well prepared nursery bed fully nourished with organic matter and fertilizers. Before sowing the soil is solarized to eliminate weeds, harmfull microbes. In solarization the beds are watered and then covered with a black polysheet for 10-15 days during hot summer months of May and June only. The bed is then prepared and seeds are sown and irrigated. Seeds germinate in 5-7 days and the plants are ready for transplanting in a months time.

Spacing : Spacing is kept as per variety. French varieties are dwarf in nature and attain a height of 15-20 cm only, where as the other varieties of African types attain a height of even up to 1-1¼ meter. Spacing for French is kept at 15 cm and for others it is kept at 60 cm.

Time of Sowing : Sowing is recommended during June to mid September and January for Summer, Rainy season and winter cycles. Flowers are available during rainy season, winters and with the beginning of summers.

Manuring and Fertilization : Y.S. Parmar University of Horticulture and Forestry recommends a dosage of 120 kg Nitrogen and 20 tons of FYM per acre for a very good crop of flowers.

Cultural Operations : To provide a bushy appearance to the plant and with more number of branches per plant for flowering, plants are pinched from the apex after 30-40 days. Growth retardants have shown very good effect in controlling the vegetative growth.

Irrigation : Irrigation is a necessary factor during summers when the frequency of irrigation is repeated after every 4-5 days

Harvesting : Fully opened flowers are harvested during the early hours of morning. Harvested flowers are immediately cooled down by putting a drenched gunny bag, which keeps the freshness. Flowers are then transported to the nearest market.

Yield : Yield of cut flowers ranges between 4-4½ tons per acre for French Marigold and this has been recorded as 8-9 tons per acre for African varieties.

CHAPTER - 10

FOLIAGE PLANTS

Another big group in the Floriculture is the Foliage plants of ornamental horticulture, which can be shrubs and climbers too. These group of plants present a very eye catching view in the Landscape gardening, as potted plants and plant arrangement providing a colour variation etc. Amongst the foliage climbers are *Ficus repens, Hedera helix, Monstera deliciosa*, Philodendrons, Scindapsus, *Vernonia elegnaefolia*. In the shrubs are the Ferns and the Cactii group of pants. In the tree and miscellaneous group the plants are Palms, Bougainvellias, Amherstia, Chorisia, Anthocephalus, Magnolia's etc.

A. FOLIAGE CLIMBERS

- ***Ficus repens :*** is a slow growing creeping plant which attaches itself to the part it gets for support with its Ivy like roots. Leaves are small dark green in colour and attach to the support providing a beautiful green cover. This cover can be designed further to enhance the beauty by putting names, different shapes of animals or birds and designs etc.
- ***Hedera helix :*** This is one of the beautiful climbers of the temperate or cool zones of the country. These climbers are generally seen climbing on the tree trunks. Leaves are triangular with marked veins. It creeps along the walls, tree trunks and is propagated by cuttings. This has very many different variations in its leaves. It loves a sunny area.
- ***Monstera deliciosa :*** Short thick stemmed plant with aerial roots coming out from the sides of the plant. It is

slow growing. Foliage of Monstera, is very deep dark in colour, with deep cuts and glossy shine. Other variations in this specie are *M. variegata* which has variegated leaves with tiny perforations and deep margin cuts.

- ***Philodendron :*** This is an evergreen climber very much similar to Monstera, produces very decorative foliage and climbs with the help of aerial roots. Foliage is deep green plain or variegated with a glossy shine.
- ***Scindapsus :*** In India this plant is very popular in every home as a climber and is known as Money Plant. It loves very low or diffused light and the vine requires proper support and direction for its proper spread. Leaves of Scindapsus are heart shaped very shiny and glossy. More species exist in this genera which are variegated in nature.
- ***Vernonia elegnaefolia :*** This is a particular plant which can be trained to grow on the window sills or on terraces. Flowers are produced in the months of February – March and are very attractive and hang like garlands. It grows vigorously and covers the desired portions immediately.

B. FERNS

These are the plants which are moisture and shade loving. Ferns species are available in every growing region whether it is tropical / sub-tropical / temperate in climate. Ferns feed best on the decayed bio-matter of the plants in the place of habitat. Most popular Ferns which are well known in the ornamental horticulture and floriculture industry are the Adiantum (Maiden hair Fern), *Asplenium nidus* (Broad leaved Fern), Lygodium Fern (Climbing fern), Platycerium Fern (Stag's Horn Fern), Nephrolepis, *Lindsea odorata* (Fragrant Ferns). Ferns are propagated by root division during the plant growth periods in the months of July, August.

- ***Adiantum*** : Genera *Adiantum adorns* in it other species like caudatum, cuneatum, decorum, macrophyllum, tenerum var. farleyense. Adiantum genus has small plants with tiny leaves attached to the stem on both sides. Species caudatum has giant leaves. *A. tenerum* var. *farleyense* makes a very good pot plant.

- ***Anemia :*** This fern is also known as Ash Fern because of its ash coloured leaves.
- ***Asplenium*** : Most of the species in this group thrive in the hills of North India. *A. nidus* is also known as Birds nest fern, since the center of the plant is so broad enough to house the nest of the bird. Fronds of this genus are very long and spatula type.
- ***Cyathea*** : This is also known as tree fern, since this has a tree like trunk which takes a number of years to develop. *C. gigantean* has been declared as a threatened plant.
- ***Davallia*** : This genus has been described as one of the handsome genus of Ferns. It has white wooly creeping stems which cling on tree's and wood logs. The other best specie of this genus is *D. bullata.*
- ***Drynaria*** : This fern is known as Oak Leaf Fern.
- ***Lygodium*** : The genus has thread like stems with a climbing nature. *L. japonicum* is the common specie suited for humid climates.
- ***Nephrolepis*** : The genus is very easy to grow and cultivate. This fern thrives best from the Himalayan ranges to the plains of India. Other species of the genus are exaltata, acuminata.
- ***Osmunda*** : *O. regalis* is known as Royal fern and is one of the most handsome ferns.
- ***Pteris*** : A common fern of the plains of India. This fern has variegation in its species of dentate, cretica.
- ***Platycerium*** : Stag Horn Fern genus has similarity to the horns of the stag.
- ***Polypodium*** : The fern is evergreen and very easy to grow. *P. vulgare* is popular species with 6-12 inch long fronds. *P. aureum* is another specie of the genus, whose fronds are deeply cut and bluish green in colour.

C. PALMS

Palms thrive well in sandy soils with warm humid climate. These plants have the capability to grow in the sun as well as in the shade too. Palms have a special place in floriculture for placing the plants to cover, decorate and layout avenues in the garden as well as in the farm houses. Palms make very well potted plant culture for commercial usage. The species which have very good demand are Kentia, Sago, Cycad. Palms are propagated from seeds, root division and suckers. Ornamental palm Species of floriculture are —

- **Feather Leaved Palms** : *Caryota urens, Chrysalidocarpus lutescens, Hyophorbe, Kentia belmoreana, Oreodoxa regia* (Royal Palm), *Phoenix roebelenii, P. rupicola.*
- **Fan-Leaved palms** : *Corypha umbraculifera, Chamaerops humilis, Lantania commersonii, Licuala grandis, Livis- tonia chinensis, Pritchardia pacifica, Raphis himilis, Thrinax argentia, Stevensonia grandiflora, Washingtonia filifera.*

D. CACTII AND OTHER SUCCULENTS

Cactii are the plants which have gained popularity because of very many exciting shapes, blooms and ease of culture and long life without least care. These have the character to thrive even under extreme conditions of water and heat stress.

ANNUAL FLOWERS

Annual flower culture is done in two seasons, summer and winter. Flower species are different for both the seasons. Annuals are planted for landscape beauty, edging and also for cut flower culture. Summer annuals are only a few in number, where as winter annuals are a good number with a variety, colour and selection to make.

Summer Annuals : Aster, Amaranthus, Balsam, Celosia, Cosmos, Coreopsis, Companula, Gomphrena, Gaillardia, Kochia, Marigold, Petunia, Morning Glory, Sunflower, Zinnia, Salvia.

Winter Annuals : Antirrhinum, Calendula, Candytuft, Cornflower, Clarkia, Carnation, Chrysanthemum, Poppy, Dianthus, Pansy, Dimorphotheca, Gazania, Godetia, Hollyhock, Helichrysum, Larksspur, Lupinus, Mesembryanthum, Nasturtium, Nemesia, Daisy, Phlox, Schizanthus, Stocks, Sweet Pea, Sweet Sultan, Allyssum, Verbena.

Nursery Raising : Nursery beds are prepared by adding well rotten FYM and giving a deep digging or ploughing to pulverize the soil for easy germination of seedlings. During summers the beds are solarized to keep these free from microbes and harmfull pests. During winters the beds are treated with 0.1% Bavistin before sowing. Seeds are sown at proper depths depending on the size of the seeds. Seeds are then covered with thin layer of sieved FYM and water is sprinkled over the bed. To keep these beds moist with proper moisture content for germination, mulch's are also provided of polysheet or grass. Seeds are generally ready for transplanting after 30 days.

Sowing Time : Sowing times are different under Indian conditions, as there are various types of climates. In the Hilly regions of North India, middle hills and plains seasons are very much different as such sowing times also differ.

Time of Sowing

Region	Summer / Rainy Season	Winter
High Hills	April – May	September- October
Middle Hills	April – May	September-February
Plains	March – April	September-October

PART - II

CONTINENTAL VEGETABLES

CHAPTER - 11

VEGETABLES

India is a unique country with various agro-climatic zones in it and possesses the potential to grow all the vegetables of the world whether these are the prevalent ones which are known as Tropical / Asiatic / Oriental, which have a developed culture of the tropics and taste of the Asiatic populace. Continental vegetables is a known name of the European countries vegetables or that of the Temperate climate cultured vegetables and a developed taste of the western world. Prior to Independence ordinary Indian farmer was unaware of continental vegetables, which has a history of its own right since the coming of Dutch and English people to the throne of India. These vegetables had the taste of the European and the western world the Dutch, French, Spanish, Portugese.

HISTORY AND INTRODUCTION

Culture of continental vegetables in India has been introduced long ago in the 17th century but its prominence could not be observed till the middle of the 20th century. Oldest pockets of its cultivation are Hills of Ooty, Kodaikanal in South India, Darjeeling, Kalimpong in North Eastern region of India and Shimla Hill areas of the Northern India. Commercial cultivation of continental vegetables can be rightly attributed to the efforts and encouragement of Mrs. Campbell, Diplomat in U.K. High Commission. Broccoli, Chinese cabbage, Red Cabbage, Brussels sprouts in the Cole crop group, Leeks, Chives, Shallots in the Liliaceae were the main crops cultivated then. No cuisine in the English dish serving, was ever complete without Parsley, Celery and Lettuce forming a part of it. With this culinary herbs also came into cultivation in these areas during 1851. Rosemary, Oregano, Marjoram used to be grown in backyards of gardens. The cultivation of each crop was restricted to its local use by individual families and never formed a part of larger open cultivation by farmers. In Kashmir, history of continental vegetable culture can be traced back to 1931, when commercial cultivation of Asparagus was taken up.

After 1947, when India attained freedom, gates of the country were opened for all the countries of the world through their chanceries/embassies/missions/consulates, collaborations in Industrial, educational, engineering spheres for major economic developments. Apart from this tourist traffic to India increased. Thereby demand for continental vegetables came in focus with a growth in demand from all these user segments from the western world. Initially the cultivation of these vegetables, culinary herbs was restricted only to industrialist farmers in Delhi, Calcutta up to 1960. But in the decades of sixties coming up of 5-Star culture and Hotels in good numbers in the cosmopolitan cities of the country, increased tourist traffic, increased global populace in the capital and Indian cities, changing life styles of Indians, demand for continental vegetables increased day by day. With the sale of these vegetables coming into the markets of Delhi, Calcutta, Bombay and Madras and offering 20-25 times higher prices than the local Indian vegetables, continental vegetables grew in popularity, both on the table and in the cultivation also. In the decade of 1980 the emphasis shifted for cultivation of these vegetables to the local farmers. For regular supply of continental vegetables to the markets and Hotels throughout the year and catering to the demands of the metros of India, not only the area under cultivation increased but also more number of farmers took the culture in their own lands. Credit of market sales of continental vegetables popularity in Delhi can be given to Allied Fruit and Florists Khan Market New Delhi, who encouraged the culture and sale of these vegetables counter, which was newly started then in the year 1961. Owner Sh. Panna Lal, who was very young then tells about huge take off sales by the foreign. Panna Lal then encouraged the cultivation of these vegetables in Delhi conditions.

PRESENT STATUS

Only a few decades back when these vegetables were being considered by the elite and all as luxury vegetables of the Europeans, the same has now come to stay in India as the most demanding and desired vegetables at the retail shelf. Continental vegetables in the period prior to 19th century had only restricted culture in the home gardens of the English, Dutch, Spanish people in India. At that time only vegetables of the Asiatic or nearby origin centers were available in the market and the number was also restricted. However today the markets, shops and the trading centers are full of a huge range of vegetables with their different cultivars pouring in with different colours and adding a variety and taste to the recipes and dish's of the common man too. This spurt in the vegetable markets of the country was observed immediately after fifties but, remained steady and slow then. With the Tourism increase and Diplomats from all over the world pouring in every part of the Indian metros raised the demand and created an awareness in the mind of the farmer for this growing demand of new type of vegetables with very high prices.

Delhi is the receiving center for 70% continental vegetables produced from South India and North Eastern region, because of very good support prices inter-alia high farm gate returns and high demand in the market. From Delhi the same vegetables are shipped to centers in Mumbai, Kolkata and Chennai. Remaining 30% production of these areas is routed to these metros directly but only through selective centers without any regular marketing base for these vegetables. Nainital in Uttaranchal, Shimla in Himachal Pradesh and Srinagar in Jammu and Kashmir has been the late entrants in the culture of continental vegetables. These hill stations sprung up only because of market catering for the summer demands of the metros. Continental vegetables are otherwise widely cultivated during winters in the plains of India adjacent to these metros.

There is no regulated market in Delhi or else where in the country which controls the rates, trade of these continental vegetables and herbs. Opening up a separate regulated market or amalgamating the sale in the Azadpur's Agriculture Market's New Sabzi Mandi and making aware the auctioneer's are the other viable alternatives to cater to this growing demand, which has become of prime necessity now. Continental vegetables as of now are directly marketed and sold to retailers in Delhi and other cities at very high prices. Retail collectors in Delhi then cater to supplies of their dealer network in the trade for these vegetables. This has been necessitated because of regular demand and catering. In Delhi the major business of retailing of continental vegetables is controlled by two markets in NDMC area. These are Khan market and INA market. The other pockets of retail sale and collection in Delhi are Bengali market, Connaught Palace, Defence Colony, Greater Kailash, Vasant Vihar. There are 3 traders in Khan market, 5 in INA market and one each in other markets of Delhi. During winters the supply of vegetables to these market outlets is maintained from Sonepat, Gurgaon, Delhi Farms and from places as far like Amritsar and Jullundhar. Production, total sale figures for both the seasons are not available but the increasing market demand is the replica of the increased consumption every year rising and Indian society also getting tuned to the continental vegetable delicacies and changing tastes. Acceptance of Broccoli, Leeks, Shallots, Button Mushrooms in the regulated market of Delhi as regular vegetables unlike other Indian vegetables is a good indicator of increased demand for these continental crops. Introduction of Fresh Herb sales in these markets has become a regular feature of the growing demand for herbs.

Table: Showing Vegetable Area and Production in selected Indian States

State	Area in Ha. Production in MT					
	Area			Production		
	1993-94	1994-95	1995-96	1993-94	1994-95	1995-96
Arunachal Pradesh	17217	17217	17217	80117	80117	80117
Himachal Pradesh	38325	38369	38869	537800	544000	569000
Jammu and Kashmir	32900	32900	32900	353706	353706	353706
Meghalaya	25436	25400	22292	238071	199845	211452
Sikkim	7544	8693	8703	46911	28792	50640
Tamil Nadu	170922	174818	174818	4389043	4397932	4397932
Uttaranchal	84000	86503	88092	717850	774408	790530
West Bengal	470300	490000	510000	4858500	5340000	5391000

Source : NHB- Indian Horticulture Database – 1998

Table : Showing Vegetable wise Area in India

Area in Ha.
Production in MT.

State	Area			Production		
	1993-94	1994-95	1995-96	1993-94	1994-95	1995-96
Brinjal	300703	420091	434202	4612231	6232254	6443062
Cabbage	23681	216759	218381	3592922	3906264	3861684
Cauliflower	188545	216142	220025	2872942	3244099	2472987
Okra	295227	415890	430525	3029453	3988765	4031811
Onion	367500	378600	395500	4006400	4036100	4080000
Peas	181611	218226	223965	1528371	2306298	2341313
Potato	1047100	1069400	1109000	17392400	17401300	18843300
Tomato	345943	351787	355684	4933980	5261327	5441969
Others	1917623	1726077	1948165	23818014	20909510	24077438
Total	4875933	5012972	5335447	65786713	67285917	71594564

Source : NHB – Indian Horticulture Database - 1998

Table : Showing Vegetable Area and Production in Selected Indian States

Area in Ha.
Production in MT.

State	Area			Production		
	1993-94	1994-95	1995-96	1993-94	1994-95	1995-96
Arunachal Pradesh	17217	17217	17217	80117	80117	80117
Himachal Pradesh	38325	38369	38869	537800	544000	569000
Jammu and Kashmir	32900	32900	32900	353706	353706	353706
Meghalaya	25436	25400	22292	238071	199845	211452
Sikkim	7544	8693	8703	46911	28792	50640
Tamilnadu	170922	174818	174818	4389043	4397932	4397932
Uttaranchal	84000	86503	88092	717850	774408	790530
West Bengal	470300	490000	510000	4858500	5340000	5391000

Source : NHB- Indian Horticulture Database – 1998

Table : Showing Vegetable Production of World Countries from 1989 – 1996

Area in Ha.
Production in 1000 MT.

Country	1989 – 91	1994	1995	1996
China	128149	189870	201825	202155
India	59321	64614	64671	64672
USA	30842	35189	34162	34393
Turkey	17963	19645	21731	20796
Japan	14457	13419	13589	13589
Italy	14436	13883	13555	13555
Russian Federation	—	10079	11551	11099
Korean Republic	9768	10128	10759	10562
World	468518	545015	566367	565523

Source : NHB – Indian Horticulture Database - 1998

CHAPTER - 12

CONTINENTAL VEGETABLES

CONTINENTAL VEGETABLES CLIMATE AND CULTURE

Vegetables are cultivated all over the world in two different seasons i.e. winter and summer. European continent with its geographical location, climate falls in the temperate and sub-temperate group of crop culture, where only crops loving a cold climate are cultivated. These crops in Europe are cultivated from March-October. The very crops which are under culture in this region are called continental/English and other crops being imported to Europe from the tropical world regions are known as Exotic or Oriental or Tropical or Asiatic crops. Also those crops of the European region when introduced by European settlers in India right since 16th Century came to be known and reckoned as continental vegetable crops. Main vegetable and herbal continental crops of importance, group wise belong to Brassica's, Liliaceae, Umbellifereae, Leguminoseae and Labiatae. These vegetables are prized in the western countries because of multiple usage in the cuisine as vegetables, dressings, decoration, garnishing, salad preparations, desserts, sauces and chutneys etc. Continental vegetables has also found usage in Chinese cookings.

Oriental/Exotic/Tropical Vegetables: Oriental/Exotic/Tropical vegetables under culture in the tropical region of Asia and finding appeal with the European population and the ethnic immigrants from the Asian region found a very good response in the western world. Amongst this group which is very much based on the Thermo classification of vegetables, vegetables like Okra, some members of the courgettes / marrow group are in very high demand in the EU markets. Vegetables in the Continental and Oriental groups are :

CONTINENTAL VEGETABLES	INDIAN VEGETABLES
COLE CROPS	
• Broccoli	Cauliflower
• Red Cabbage	Cabbage
• Brussels sprouts	
• Knol-khol	
• Kale	
BRASSICAS	
• Chinese cabbage	
LEGUMINOSEAE	
• Snow Peas	Shell Peas
• Fava Beans	Broad Beans
• Lima Beans	Sem Phali
• Guar Phali	
LETTUCES	
• Romain	
• Butterhead	
• Iceberg	
• Cos	
• Endive	
COURGETTES AND MARROWS	
• Zucchini	Pumpkin
• Gherkins / Cornichons	Squash's
• Cucumbers	
• Karela	
• Bhat Karela	
• Kakri / Tar	
• Luffa	
• Spongegourd	
OTHERS	
• Asparagus	Radish
• Globe Artichoke	
• Jerusalem Artichoke	Turnip
• Eggplant	Brinjal / Eggplant
• Rhubarb, Swiss Chard	
• Baby corn	

- Parsnip
- Chicory
- Cherry Tomato
- Collards
- Water cress
- Pak Choi
- Salsify
- Horse Radish
- Avocados
- Chayote

Tomato

Okra / Ladyfinger

GREEN'S

Mustard

Fennugreek

Fennel

Amaranth

Spinach

HERBS

- Celery
- Parsley
- Fennel
- Sweet Basil
- Leeks
- Shallots
- Chives
- Rosemary
- Thyme
- Oregano
- Marjoram

Coriander

Mints

COLE AND BRASSICA'S : Crops of this group include, Broccoli, Brussels sprouts, Cabbages (Green, Red, Savoy), Kale, Knol-knol and Chinese cabbage.

- **Broccoli :** This is *Brassica oleraceae* var. *italica* or sprouting Broccoli. Its heads are similar to Cauliflower, but are green or pink in colour. Pink Broccoli is the most hardy vegetables of winter coles, withstanding frost and winter damage, but takes 18 –20 weeks to maturity and is low in demand. Broccoli with green heads is much in demand and takes 8-9 weeks for maturity of heads.

- **Brussels Sprouts :** This is the second most popular vegetable of the Cole's and Brassica's. It is *Brassica oleraceae* var. *gemmifera,* originated in Belgium. The popularity of this vegetable is due to longer picking span and withstanding of severe winter colds. Under Indian conditions when it is to be grown as main crop the cropping can be extended up to May month in the plains. This can be compared to green cabbage but with mini heads arising in the axil of leaf in the main stem.
- **Red Cabbage :** This is another member of the group and is similar to green cabbage but with a colour variation of the leaves and the head down to the incore inside. This is used both as a salad, dressing and garnish.
- **Chinese Cabbage :** This is also green Cabbage. Head is less compact than Green Cabbage. Its main usage is in the Chinese cookings.

LETTUCES : This group of the greens is generally deployed for green servings of the heads. Lettuces are of four main types i.e. Heading, COs types, Loose leaf and Crisp Head. Various specific plants of this are —

- **Romain :** These heads are not crispy but open in form. Leaves are not allowed to mature. Heads are harvested immature. Slower to bolt than other Lettuces.
- **Butterhead :** This is another variation of the Lettuce heads where the head is silky in touch and loosely compact, hence termed as Butter Head.
- **Crisp Head :** ICEBERG lettuce is very crispy in texture and commands a high market demand.
- **Leaf Lettuce :** This is also known as home garden Lettuce grown mostly in Green houses in Europe and else where.
- **Endives :** This is also like Romain type Lettuce. Leaves are dentate type and open. Central part of the rosette is blanched to create a variation of the quality for salads. Leaves are much sweeter and tastier. Endives have another variation from Witloof Chicory, to make Chicoons from the roots, for salads.

COURGETTES AND MARROWS : This group in the Cucurbits has Zucchini and Gherkins which are the developed immature fruits from the flowers.

- **Zuchhini :** This is a cylindrical fruit of the pumpkins with range of light to dark green colour of the skin. This Marrow possesses an added aroma feature in it. Prior to 1986 demand for Zuchhini was very negligible in the Indian market. During the year 1988 its commercial culture was undertaken by Nijjer Foods Ltd. and since then the demands is on the rise and has outstripped supplies in the market. Fruits are harvested immature to be taken along with the skin. Development of the fruit from flowering takes only 8-10 days for harvest maturity.
- **Gherkins :** These are the immature cucumbers of the size 2-3 inches long. Fruits are harvested within 3-4 days after flowering. Gherkins are used in salads and as processed form as pickles in vinegar.

ASPARAGUS : The Asparagus vegetable is the stem part of the plant species *Asparagus officinalis* which appears in the form of spears during spring season. These spears come up from the root crown and develop very fast. These are harvested just below the soil with sharp knifes within 2-3 days of appearance. Fiber in the spears is not allowed to develop which spoils the quality of the spears. Asparagus spears are used both as green and in canned forms. Asparagus is one of the prized vegetables in the continental group comprising a very high demand in the world as an export crop. Countries like Thailand, Taiwan, Israel has mastered the technique of Asparagus culture so much, that the spears can be made available to the market through out the year and thereby earning a good return in comparison to the main season crop.

ARTICHOKES : Artichokes are of two types, the Globe and the Jerusalem. Globe Artichoke develops from the immature flower where as Jerusalem Artichoke is the development of the rhizome.

BABY CORN : Baby corn is the latest development of the decade of sixties and has come up to be regarded as one of the prized vegetable of the continental or Exotic group. This was developed by Thailand and now commands as one of the export vegetable for very many countries of African continent. Kenya, Zambia, Zimbabwe has earned a name for Baby Corn trade for export into the EU. Largest exporter to the EU is Thailand. Cobs are harvested immature when the silk has just emerged and the size of the cobs is 4 – 5 inch long.

HERBS : Following herbs are much in demand for regular usage in the Hotels, Restaurants and by the Caterers.

- **Celery :** Celery is *Apium graveolens* and locally known as Ajmod in vernacular in India. Both leaves and stem are useful as garnishing. Seeds of celery yield an oil known as Celery oil. Celery is best friend of mankind having numerous usages as a digestive and circulatory tonic.
- **Parsley :** This can be compared to Indian coriander. Parsley is of three types i.e. French, Italian and Curled. Curled Parsley is crisp, tightly curled foliage is used in garnishing and decoration. Italian Parsley has strong flavour and aroma.
- **Rosemary :** Needle like leaves are the actual product of the plant in demand in the market. These are available in the dried form in 50 gm package. Herb is quite high in demand for Non-vegetarian dish's, in enhancing the flavour and aroma of the recipe.
- **Sage :** Sage is also a dried herb available in the market and is well suited for Meat recipe's and dish's.
- **Thyme :** These are plants with 3 different flavours and aromas. Leaves are dried and used both fresh and dried.
- **Marjoram :** Both Marjoram and Oregano are covered in this group and are relished for their inclusion in sauces, marinades etc.

In the grouping of continental vegetables these can be further classified as low and high demand crops. Swiss Chard, Parsnip, Salsify, Rhubarb, Chicory, Horse Radish crops which are very small in demand. This division has been necessitated because of the fact, that high demand vegetables are in demand throughout the year and in very good quantities and possessing export potential too. In the present times coming up of different Import / Export trade regions in the world, access to these is very competitive and rewarding with ensured high returns for the fresh farm produce.

Survey of low demand vegetables, effected for USAID during the year 1991 February to March of Embassies, High Commissions, Hotels and Hotel Chains, Restaurants, Caterers, reveal that with in a time span of 5-7 years the demand for low demand group of vegetables can eclipse the demand of other vegetables. Likewise culinary herbs has two facets of demand both as culinary herbs for dressings, sauces, mayonnaise etc. and as essential oil bearing crops. Cultivation of these can be undertaken in a modest way viewing the local demands and export potentials for dried herbs and oils.

Continental Vegetables Rich in Nutrition : Continental Vegetables are a rich source of proteins, minerals, amino acids and vital elements like Calcium, Phosphorous. Vegetables like Broccoli, Artichokes, Brussels sprouts, Celery, Parsley are highly rich sources of proteins, minerals and elements in comparison to Indian vegetables like Cauliflower, cabbage. Emphasis of the Government is very much aimed at restoring nutritional health standards in the masses so that every Indian is nutritionally healthy and fit and there is no cause to suffer from disorders due to malnutrition from protein deficiency, energy and Iron which are the causes to anaemic developments in the body and blindness. Continental vegetables being rich in these components are thus capable of providing 3 times more energy than Indian vegetables.

Table : Showing All India Demand of Culinary Herbs and Prices / Kg.

Herb	Estimated yearly demand in kg	Price/kg to farmer in Rupees	Consumer price in Rupees
Oregano	500	150	300
Marjoram	250	150	300
Sweet Basil	1000	150	300
Thyme	200	250	400
Terragon	100	150	300
Rosemary	1000	300	600
Sage	200	250	500

Source : USAID Report – 1991, Continental vegetable crop culture in HP.

The data projected on demand and prices above is a survey from the Delhi markets, based on the actual prices which were available to the grower and the consumer then, without any developed packaging of the dried herbs. With a standard packaging development for the herbs and placing this at par to world standards in the markets with attractive cartons, jars, bottles, price accrual for the same amount of herb can be enhanced further to 5-8 times from the presented prices.

Market Potential for Continental Vegetables : These vegetables of the west have opened up vistas for not only meeting the local demands but also exposing the economy of India for entering into the export arena. Continental vegetables are much in demand these days, because every consumer prefers a fresh product. Also post harvest technology has made the process of keeping and maintaining the freshness of vegetables more simpler and ensuring regular availability of the product. In Europe other than the controlled temperature Green House culture brings up very many crops during the year but the continent is capable of having only one crop during the year in the open system. Thus nearly every country in EU imports and re-exports these products to other countries where demand exists. One such market is the Gulf Market in the Middle East which caters to the import and export requirements of other countries. Asparagus, Artichokes, Brussels sprouts, Baby corn, Avocados in the exotic group of vegetables, Ginger, Garlic, Dried Herbs are in continuous and increased demand. Potential of India lies in its weather and location of the continent in the middle of all exporting countries of South East, Far East and Pacific for their crops. India has the capability of producing crops throughout the year and meeting all the commitments of the Importers in the Gulf and European Union markets.

CHAPTER - 13

VEGETABLE CLASSIFICATION

INTRODUCTION

Vegetable is a herbaceous plant or a portion of plant, that is eaten whole or in part as vegetable as raw or cooked. In the Botanical classification there is a distinct division between a vegetable and fruits. However botanical terminology qualifies a vegetable to be a fruit also. Tomato, Beans, Pepper, Cantaloupe are fruits. Fruit in the strictest terms is a ripened ovary containing seeds with its adjacent parts eaten at maturity. Fruits are sweet too, which is seldom in vegetables. Vegetables are high value crops having all the basic characteristics for their classification on the morphological and agronomical characteristics. Vegetables have found classification under very many groups of Thermo sensitivity, Botanical, Usage, Edible parts and on the agronomical side for salt tolerance, root depth etc. Widely accepted classification of vegetables is Thermo classification, which places the vegetables into two clear groups of cold and hot season growing crops. Continental vegetables fall under the Cool season and Oriental vegetable crops under the Warm season crops. Both groups have vegetables of variable adaptations to temperatures from low, medium to high temperatures which have been placed as under.

Botanical Classification : Botanical system characterizes each vegetable plant on the basis of similarities in its morphological characters of flower, stem, leaf, root, habit of growth bush, twiner or a vine. The other vital classification criteria in the botanical system includes the Community, Division (Angiosperms/ Gymnosperms/Phanerogams), Class (Dicot/Monocot), Genus, Species and lastly the Botanical group. Vegetables as such, deserve rightly to be classified botanically, but this classification is only for the research scientist and students. However, for the farmers and other practical groups culturing vegetables in the fields, gardens and farms, other classifications are the most suited for better understanding of the nature of the vegetables. Some vegetables like Mushrooms which belong to the Fungi of Cryptogams do not fit into the Spermatophyte community of plants. Botanical classification thus finds a general description based on the Class and the Family. In this Monocots include Allium, Dioscorea, *Zea mays*, Asparagus from the Alliaceae, Dioscoriaceae, Poaceae and Liliaceae families. Most of the vegetables fall in the second group of Dicots covering families of Brassicaceae, Convolvulaceae, Asteraceae, Chenopodiaceae, Leguminoseae, Cucurbitaceae etc.

CLASSIFICATION SYSTEMS

I. Thermo Classification of Vegetables

A. COOL SEASON CROPS

Group	Preferable Av. Monthly Temperature	Intolerant Temperature	Crops	Remarks
A	15.8-18.3°C	21.1-23.9°C	Spinach, Cabbage, Parsnip, Broccoli, Radish, Beet, Turnip, Rutabaga, Cauliflower	Slightly tolerant of freezing
B	15.6-18.3°C	21.1-23.9°C	Lettuce, Celery, Artic-hoke, Endive, Mustard, Carnot, Chard	Damaged near maturity by freezing
C	12.8-23.9°C	-	Asparagus, Garlic, Kale, Brussels, Sprouts	Tolerant of frost
B. WARM SEASON CROPS				
D.	18.3-29.4°C	-	Sweet corn, Pepper, Snap Bean, Squash Pumpkin, Lima Bean, Cucumber Tomato	Intolerant of frost
E	Above 21.1°C	-	Muskmelon, Watermelon, Sweet Potato, Eggplant, Okra	Long season crops

II. Usage and Botanical Classification

Based on the Botany of the vegetable crops and their usage, vegetable classification has thus been done as Pot herbs/Greens, Salad, Cole, Root, bulb crops, pulses, members of Cucurbit, Solanaceae family.

- **Pot Herbs / Greens** : Spinach, Swiss Chard, Newzealand Spinach, Mustard, Collards, Dandelion, Fennugreek, Purslane, Rhubarb, Water cress, Kales, Fennel, Colocassia, Chenopodium.
- **Salad Crops** : Celery, Chicory, Lettuce, Watercress, Endive.
- **Cole Crops** : Cabbage, Brussels sprouts, Cauliflower, Kohlrabi, Broccoli.
- **Root Crops** : Crops with fleshy roots included in this group are Beet, Turnip, Carrot, Rutabaga, Parsnip, Radish, Salsify, Celeriac.
- **Bulb Crops** : Onion, Garlic, Leek, Shallot, Welsh Onion, Chives.
- **Tuber Crops** : Potato.
- **Pulses** : Shell Peas, Snow Peas, Beans.
- **Cucurbits:** Cucumber, Pumpkin, Muskmelon, Squash, Watermelon, Bitter Gourd (Karela), Chayote, Kakri, Bhat Karela, Ashgourd, Luffa Gourd.
- **Solanaceous Fruits** : Tomato, Eggplant (Brinjal), Capsicum / Paprika / Pepper.

III. Edible Part Based Classification

This classification has identified all the sub groups based on the plant tissue i.e. Roots, Stem, Flower, Leaf, Immature flowers, Immature and mature fruits. The best suited classification of vegetables is on the basis of the edible parts which have become most relevant in the present context and is a very easy way to understand the vegetables in totality and their usage.

- **Roots** : Roots can further be sub-divided in two groups of - Enlarged tap root and Lateral roots. Different vegetable crops in both the groups are – **Enlarged Tap Root** "Beet, Rutabaga, Carrot, Turnip, Radish, Parsnip, Salsify,

Celeriac, Horse Radish. "In the **Enlarged Lateral Roots** vegetables included are "Sweet Potato, Cassava".

- **Stem :** Stems have another classification having a clear distinction with starchy and non starch and stems, above and below soil. In this classification vegetable plants included above soil are - Asparagus, Kohlrabi/Knolkhol, Tree Fern for Non starchy stems. For stems below ground Potato, Jerusalem Artichoke, Yams, Taro are vegetables containing starch.
- **Leaf :** Leaves are used for **Salads, Potherbs** and other **miscellaneous** uses. Plants categorized in the **miscellaneous** group are from Alliaceae family of monocots with Onion, Garlic, Leek, Chives, Shallot. In the Broad leaved group leaves used for **Salads** are of Lettuce, Chicory, Endive/Escarole, Cabbage, Celery, Chinese Cabbage, Kale, Brussels sprouts. As **Potherbs** for cooking, the group includes Spinach, Newzealand Spinach, Kale, Chicory, Chard, Amaranth, Chenopodium, Chinese cabbage, Mustard, Fennugreek, Fennel, Dandelion, Cardoon (Petiole only), Rhubarb (Petiole only), Water cress, Taro Leaves, Collards.
- **Flowers :** Flowers are used both immature and mature in this classification. **Immature Flowers** group vegetables are Cauliflower, Broccoli, Globe Artichoke. In the **Mature** group of flowers used are for salads and includes flowers of Nasturtium, Bauhenia variegata, Calendula, Viola and Roses.
- **Fruits :** In this group categorization has further been done as Immature and Mature fruits. In the **Immature Fruit** group, fruits of Peas, Chayote, Okra, Snap Beans, Faba Beans/Broad Beans, Summer squash, Sweet corn, Lima Bean, Cucumber, Eggplant (Brinjal), Kakri, Karela, Bhaat Karela, Luffa's. In the MATURE FRUIT group, these are fruits from the Gourd family – Pumpkin, Winter squash, Muskmelon, Watermelon and Capsicum, Pepper, Tomato from family Solanaceae.
- **Seeds :** This group includes Coriander, Kalaonji, Fennel, Zeera, Dill and Ajowain seeds, Beans, Pea.
- **Rhizomes :** This includes in it the crops of Turmeric and Ginger.

IV. Classification on Salt Tolerance Limits

This classification stands for salt tolerance limits in the soil based on the salt concentration on saturated extracts of soils and has been classified as High, Medium, Low salt tolerance and other vegetables has the capacity to bear a tolerance limit of 1900 ppm.

- **High Salt Tolerance :** limit is that of 7700 ppm and the crops in this group are Garden Beets, Kale, Asparagus, Spinach.
- **Medium Salt Tolerance :** limit is 6400 ppm and the crops included are Tomato, Broccoli, Cabbage, Peppers, Cauliflower, Lettuce, Sweetcorn, Potato, Carrot, Onion, Peas, Squash, Cucumber, Cantaloupe.
- **Low Salt Tolerance :** limit is that of 2600 ppm and the crops included in this are RADISH, green Beans.
- **All other Vegetables :** limit is 1900 ppm. This group will include all vegetables not mentioned in other groups but are capable of tolerating maximum limit of 1900 ppm and below.

V. Acidity Tolerance Classification

In the soils pH is the criteria of its being acidic or alkaline. Higher the pH more alkaline it is and lower the pH more acidic it is. Thus crops which are tolerant to highly acidic, moderately acidic and tolerant to slightly acidic conditions have been grouped.

- **Slightly Tolerant :** The pH limit in this group is from 6.8 to 6.0. Crops which are capable of bearing this acidic level are Asparagus, Celery, Muskmelon, Beet, Spinach, New-zealand Spinach, Broccoli, Chinese Cabbage, Okra, Cabbage, Leek, Onion, Cauliflower, Lettuce.
- **Moderately Tolerant :** The pH range is between 6.8 to 5.5. Crops which bear this level are Beans, Horse Radish, Pumpkin, Brussels sprouts, Kohlrabi, Radish, Carrot, Parsley, Squash, Cucumber, Pea, Tomato, Eggplant, Pepper, Turnip, Garlic.
- **Very Tolerant :** The pH range for this group is between 6.8 to 5.0. Crops like Chicory, Rhubarb, Endive, Sweet Potato, Potato, Water Melon are included.

VI. Classification by Root Depth in Soil

Crops have specific character of remaining shallow or attaining medium or deep depth in the soil. Based on this classification crops have again been classified on the depth what it attains.

- **Shallow Rooted Crops :** Crops which can grow in shallow depths of less than 3 ft. or 80 cm depth. This includes crops like Cabbage, Lettuce, Onion, Potato, Spinach, Sweet Corn.
- **Medium Depth :** Range in this is that of crops going upto a depth of 6 ft. or remaining upto a depth of 80 – 160 cm. Crops in this group are Beans, Beets, Carrot, Cucumber, Eggplant, Summer squash, Peas.
- **Deep Rooted :** Range of depth what the crops attain is above 6 ft. depth and can go to any limits. Crops in this group are Globe Artichoke, Asparagus, Sweet Potato, Tomato.

CHAPTER - 14

COMPANION CROPS

INTRODUCTION

Landscape of nature in the forests is very nicely balanced, where menace of pests and diseases is controlled naturally by diversity itself in plants and useful insects. Plants in forests do not compete for light and nutrients. Every plant in nature thus grows nicely in the companionship of one another. It is this diversity of very many plant species in one forest, nature has created to ward off these from pests and disease attacks. Natural flowers of the forest are a source of good food and shelter for spiders and beneficial insects those which thrive and parasitize on pests. It is this diversity which is also created in companion crop gardening by growing flowers as companion crops. The idea got nourished from the natural occurrences in the forests and tried in the fields by the farmers. This provided immensely heartening results in the crop culture thus making it a separate subject of application in the garden crop culture.

COMPANION CROP PLANTING

In home gardens and commercial gardening too crop plants does need tender care and continuous attention to look for nutrient needs, pest problems, water requirement for better crop production. How beautiful and fascinating it is using time tested natures benevolence in garden crop culture, which itself not only keeps crops healthy, lush green, free from pest problems, adding to the fertility status of the soil with better yields, but requiring very less care and human attention. This is the technique of combining culture of two or more crops side by side for specific purpose. This is known as companion crop planting. Methodology involved is very clear, for crops attacked by pests, we can choose companion crops which has inherent quality to repel, trap or hide these. Other companions which provide food and shelter to attract and protect beneficial insects or friends of farmer. Some companion crops grow well together, because none competes for sunshine, water and nutrients requirements of the other. Crops of Leguminoseae family are the best in providing nitrogen to their companions and also enriching soil by fixing atmospheric Nitrogen.

SOIL ENRICHMENT

All plants withdraw nutrients from the soil for their growth, but some plants return more nutrients than what they draw i.e. leguminoseae crops like pulses, beans, peas, clovers. These leguminous plants have a interdependent relationship with the nitrogen fixing bacteria (Rhizobium or Rhizobacter), which are different for different crops. These rhizobia colonize on leguminous crop plant roots absorbing 20% of plant sugars (Rodale's Successful Organic Gardening – 1994, PA-1809 USA) what they produce. This energy is utilized to capture atmospheric nitrogen and then converting it into nitrogen compounds that plants can use. Some nitrogen of this is utilized by the plant itself and rest is deposited in the form of nodules of the roots. Roots die and decompose. The nitrogen is available during the season to boost growth of plants. Green manuring especially in Punjab, Haryana, UP areas with clovers is a traditional way of culture done only to enrich the soils through plant foliage and roots after decaying. This has formed a defined place in the crop rotation follow up in the crop culture.

REPELLING ACTION

Crop plants possess inherent capabilities in them to attract, confuse and deter insect pests. The characteristic of attraction or repulsion, which is due to the production of scents or toxins. Combination of some aromatic plants in the crop system can mask the scent of companion crops. Summer Savory may help in hiding Bush Beans from pests, while Pansy is considered as a repellant for Colarado Potato beetles. Odour of Garlic crop releases deterrent aroma, that may chase away bean beetles and Potato bugs. Mint keeps away cabbage loopers from cabbage crop. Basil discourages Tomato horn worms on Tomatoes. Try pungent crop plants as an edging around garden beds.

LURING CROPS FOR INSECT PESTS

Some crop plants have an almost irresistible appeal for certain pests. Nasturtium is an excellent example of attractants because this plant is a favorite of aphids. Black Night shade *(Solanum nigrum)* is much more alluring than Potato for Colarado beetles. Attractant plants protect crops in two ways, firstly these act as decoys to allure pests away from main crops and secondly pests get concentrated on a very few plants and it is easier to control pests either by collection or destroying the plants with the pests.

MULTIPLICATION OF BENEFICIAL INSECTS

Some of the insects are the best friends of the farmer. These insects help the crops to grow either by eating the pests or parasitising on them. These insects are encouraged by growing favorite crops of these friendly insects. Dill culture attracts and encourages pests eating spiders, lacewings and parasitic wasps, which help control caterpillars on cabbage, beetles on cucumbers, aphids on lettuce. Pollen and nectar producing plants like Yarrow, Fennel, Golden-rod provide shelter and food for hungry and beneficial insects.

COMPANION GARDEN

Companion crop plants are not different in culture from main crop. These also require the same soil preparation, timely sowing, fertilizers, irrigation and care. However the difference for companion crops is the observance of performance. It is recommended to try a combination for 2-3 times to know better and arrive at a conclusive decision. It is only thus, one can make his garden more productive and maintenance is also easier. Try also companion crops which have been proved and confirmed scientifically for their companion affect on the main crop. Farmers should also make a note of quick growing weeds or vegetable crops which harbour pests. These could be very good trap crops to lure the pests away from main crop.

CHAPTER - 15

VEGETABLE FREEZING AT HOME

INTRODUCTION

Vegetable freezing at home is a new innovation of the IQF (Individual Quick Freezing) process technology, wherein the vegetables are frozen not with the quick freezing process but with slow freezing in the home freezer. This technology is fast gaining prominence in the post harvest management of vegetables and also to increase the shelf-life and for food security reasons too. Thus this technology has paved the way for reducing crop losses due to rottage and nutritional losses for deteriorating quality. In the IQF process the processed vegetables are immediately without any loss of time are frozen with Liquid Nitrogen and other processes. This has become the necessity of the time. The various processing steps involved are a must to be done before freezing of the products at home and saving the vegetables from crop losses. Prior to this it is very important to familiarize about the technology involved. This pertains to Blanching, cooling and freezing of the products. Why this is done, is the second part, wherein the enzymes or natural proteins contained in the vegetables tend to provide flavors, texture, colour, quality of nutritional value play an important role the freezing process. What is the shelf-life of frozen product and how these are quality wise fresh, wholesome and flavorful and all the more usage in daily life are the questions one should know prior to the start of the processing.

PRESENT STATUS

Availability of seasonal vegetables throughout the year in the market and their increasing demand and fast changing life and eating styles of consumer, homes are becoming a radiant fashion of the society. Hotels, Restaurants, Catering industry are the other users in big demand for these vegetables to make available varied dishes to the clientele on demand. Society and the end consumer has to pay very high prices to get all these amenities at any time of the year. Vegetables after harvest loose their moisture and quality. Refrigerators in nearly every home is not a thing of the past but a part of the necessity and the fashion today. House wives love to freeze the vegetables in their homes and use these at times when required for parties or functions or for the service of the guests. House wives are un-aware of the fact that live vegetables which are frozen without proper processing still contain enzymes, bacteria, molds and yeast which retard the growth, quality of the vegetable. Freezing is better than drying, because it preserves more nutrients and fresh flavour. Air contact leads to flavour loss or development of off flavours. Moisture loss or evaporation in the freezer leads to toughening of the tissues. Natural enzymes cause flavour, colour, texture and nutritive value changes. Freezing does slow the enzyme activity but does not stop it till these are inactivated through the process of blanching before freezing.

WHAT ARE ENZYMES

Enzymes are proteins which change the quality, nutritive value and the colour features after harvest of the vegetable. Inactivation of these enzymes and shelf life increase of the vegetables lead to the maintenance of all these. Enzymes even continue to be active in the frozen condition too and may cause vegetables to toughen or develop off-flavours and colours. Continuous freezing of Peas without blanching develops afflatoxins which create offensive taste due to these toxins. Enzymes in the Peas are capable of surviving even the coldest regimes of temperatures and affecting thus the quality of the shelled Peas by developing these Afflatoxins.

HOW TO MAINTAIN QUALITY AND SHELF LIFE

Freezing after the inactivation of the enzymes is an excellent way to preserve fresh vegetables. Quality of the frozen vegetables is dependent on the product's freshness and how that has been handled in the post harvest till it reaches the end consumer or the cold store owner or the processor. Shortest time delivery to all these is the key for good quality of the frozen vegetables. Freezing does not improve the quality but retains and maintains the already possessed. Blanching and prompt cooling are the two steps in preparing all the vegetables for freezing, except herbs and Peppers.

METHODOLOGY

For commercial or home freezing choose fresh, young, tender vegetables when these are available in abundance and are cheap in the market. Remove dirt, dust, pesticidal residues by giving a proper wash in cold water. For maintaining a very high quality washing can be undertaken with organic wash also and then washing in clear water. Next step is the preparation of vegetables for blanching which is done by peeling, trimming, cutting, dicing. Vegetables are then blanched by boiling in hot water or steaming for a specified interval of time which will in-activate the enzymes. For maintaining original green colour of Peas etc. a pinch of common salt is added to the boiling water. Lower vegetables in the boiling water either in a wire mesh basket or cheese/muslin cloth or can be done even without all these. Start noting the blanching time immediately after it starts boiling again. Prolonged blanching period leads to the loss of vitamins, minerals and flavour. Immediately after blanching the vegetables are drained and cooled down in running cold water. Extra water is removed either by rinsing or letting it stand for some time. Vegetables are now ready for freezing.

PACKAGING

For packaging of frozen vegetables there are two distinct modes by which one can pack the processed vegetables. Frozen vegetables have a shelf-life of 10-12 months. Longer storage will not make these unfit for usage. These are ready to cook products without thawing and take very less time for cooking. These processes are —

- Dry Pack
- Open or Tray pack

Dry Pack : Pack vegetables in moisture and vapour proof packaging materials. These are polyethylene bags. Remove air from the bags by pressing and then sealing. Process of sealing is very easy in self sealing bags. Place these bags in the deep freezer, which will take 3-4 hours for process completion.

Open or Tray Pack : Place vegetables so obtained after cooling and rinsing in a tray and place it in the deep freezer for 3-4 hours, by which time freezing process is complete. Take out the tray and turn it over a clean white paper or news paper sheet spread over the table. Press the tray very slightly on the back, all the frozen vegetables will be on the paper or sheet. Fill these in poly or self sealing packs and seal. Place and stack in the deep freezer.

VEGETABLES RECOMMENDED FOR FREEZING

Asparagus, Beans of all types whether green or yellow, podded or Lima, Beets, Broccoli, Brussels sprouts, Brinjals, Cabbage, Cauliflower, Carrots, Sweet Corn, Peas, Peppers, Pumpkins, Squash, Zucchini, Okra etc. are the main vegetables which can be frozen and taste can be enjoyed beyond seasons at will. The process has the full applicability for commercial exploitation too and as a growing village industry of the future. This is an indirect boon for the farmers too, to save the post harvest crop losses due to quality, pest and diseases and add to productivity.

PROCESS

Equipment and processing tools required for the process are available in each and every village and home. These are a Gas stove, SS containers with lid for boiling water, Clean or Tap water, Cheese / Muslin cloth, Knives, cutters, cutting boards, Trays, Ice cold water, Poly packs, self sealing packs, clean paper or news paper sheets, common salt, watch for noting time, Freezer or Fridge with a very good storage capacity. The efficacy of the process depends on few facts, need to be considered before processing. these are —

- Quality of the frozen vegetables depends on the quality of the fresh produce.
- Blanching and prompt cooling are essential in preparing most vegetables for freezing except herbs and green peppers.
- Blanch vegetables in boiling water or steam. There are two basic packaging methods of the frozen product and these are (1) Dry Pack (2) Open or Tray Pack.

- Vegetables maintain high quality for 10-12 months period without any deterioration in quality.

Washing : Subject all the vegetables to be processed to clean or tap water washing, to remove dirt, dust etc from the vegetables. The process can further be more up-graded by washing the vegetables first in organic wash (a powder of purely plant origin) and then in clean water.

Trimming/Cutting/Dicing/Cubing : vegetable to be frozen is accordingly put to trimming, cutting and dicing or cubing so as to create uniformity in product appearance. The process is different for each and every vegetable as explained separately.

Blanching : Boil water in stainless steel utensil 2 times more than the vegetable weight. Let the water boil and then immerse all the vegetable contained either in a cheese/muslin cloth or wire mesh sieve basket in the boiling water and note the time immediately after it starts boiling again. After the blanching time is over bring out the cheese cloth/muslin cloth or wire mesh sieve and put it to cooling in ice cold water or running water for few minutes, till it gets properly cooled.

Draining and Rinsing : After cooling the vegetables are drained for excess water and placed on a white paper or news paper sheet.

Freezing : Freezing is the final step and is done either as a Dry pack or Open / Tray pack method.

SELECTION AND PREPARATION OF VEGETABLES FOR FREEZING

Asparagus : Select young and tender compact Asparagus shoot tips. Cut tough ends and wash Asparagus shoots to remove any dirt. Now cut, make 2-2½ inch long pieces and grade these for soft and thick shoots for giving different blanching times. For the soft grade blanch for 3½ minutes and for the thick grade blanch for 4 ½ minutes. Drain and cool the contents and freeze.

Beans : Select young tender stringless Beans. Cut the petiole and the tip and then wash thoroughly. Cut the Beans into 1 inch long pieces. Blanch these for 3-4 minutes and then drain the water and cool it.

In the case of Beans with strings cut the tips, petiole and remove the string. The cut Beans are ready for blanching and freezing process. Blanch these for 3-5 minutes, in which time the

Beans loose their enzymes and are in 50-60% half cooked condition.

Broccoli : Select tender dark green heads with stalks. To be sure there are no insect pests in the Broccoli heads immerse these heads in water containing 2% salt for 30 minutes. Remove the head, wash. Cut the stalks from the main head and peel and make into small 1 inch cubes. Cut the head into desired size pieces. Broccoli is ready for blanching. Blanch Broccoli for 4-5 minutes in boiling water.

Brussels Sprouts : Select green, compact and firm Brussels sprout heads. Remove outer leaves, trim and wash. Grade into 3/4, 1 inch and over size heads and keep these separately. Blanch these for 5-7 minutes depending on the size and grade of each, giving lowest time for smaller size and higher for the bigger size.

Brinjal / Eggplant : Select Brinjals without seeds or having immature seeds. Do not process Brinjals having brown seeds in the fruit. Wash and cut the Brinjals in the size as one desires either longitudinally or in cubes. Put salt to make a 1% blanch solution or water. 1% conforms to addition of 1 gm salt to one liter of water or one TS of salt to 5 liter of water. Blanch the prepared Eggplant for 3-4 minutes, drain and cool the contents.

Cabbage : Remove the outer hard, damaged leaves. Wash the head in water thoroughly. Cut head into medium or coarse shreds or cuts. Cabbage is ready for blanching. Blanch the shreds for 2 ½ minutes in boiling water, then drain it and cool.

Carrots : There are two types of carrots available in the market. The one with a yellow core and the other without any core are the English carrot variety. Select red, tender, un-damaged carrots. Cut and trim both the ends with a very fine knife. Braise the skin. Wash the carrots in clean tap water to remove the braised skin if any. Cut longitudinally from the center and remove the yellow core and keep the carrots as such in case of English. Cut the carrots 1-1½ cm long and now make 1-1½ cm cubes. The carrots are thus ready for freezing.

Cauliflower : Select very compact white colour heads from the market. Process of cutting, trimming, making pieces is the same as in broccoli. Blanch the cauliflower so prepared for 4 minutes.

Ladyfinger / Okra / Bhindi : Selection of Ladyfinger / Okra/ Bhindi fruits in the market is the sole criteria for the quality frozen product. Select very tender, fresh, green 2 – 2½ long fruits. Length of fruit more than 3 inch is not good for freezing. Cut the tip and the stalk cap and wash the fruits in cold running water. Blanch the fruits for 2 – 2½ minutes in boiling water.

Snowpeas : Cut the tip and stalk end of the pea and wash in cold water. Blanch the pods in 1% salt water for 2½ minutes. Drain, cool and freeze.

Shell Peas : Select shelled type of fresh green pods. Pea grains are soft and sweet in these pods. Reject mature Pea pods which are whitish green or very light green in colour and the Peas are matured non sweet. Shell out the peas from these pods and blanch in water for 4 minutes. Add a pinch of salt to bring out the shining green colour of the Peas. Drain and cool, freeze.

CHAPTER - 16

PACKAGE OF PRACTICES

INPUTS FOR CONTINENTAL VEGETABLE CULTURE

Continental vegetable culture is not different from the Indian vegetable culture. However these vegetables have been observed to be much more responsive for quality management and inputs used in the package practices of cultivation. These inputs can thus be put into two categories.

1. Critical Inputs
2. Non Critical Inputs

Critical Inputs : Inputs which are entirely utmost necessary for the best crop culture of these vegetables come under seed quality, manures, irrigation, management practices and post harvest handling, storage and marketing. Seed quality is dependent on the true to quality and purity of the seed whether hybrid or self/open pollinated pure selection. Indian farmer does not have the access to the quality seed production of these continental vegetables from the local market and has thus to rely on the imports from European or American seed companies. Seeds of the crops which have acclimatized to the Indian conditions are detailed in the package of practices for each crop. Manurial applications is the addition of the nutrients for the crop growth. Organic manures are the one's from the animal or plant origin duly decomposed into applicable form as FYM manure from decomposed cow dung, leaf or agro waste composts, poultry manure, de-oiled cakes, sewage sludge, leaf-mold etc. These are capable of supplementing the micro-nutrient demands of the crops. Green manure crops in the rotation add to the nitrogen requirement of the vegetables. These are highly good supplements of the critical input of manures as natural feeding is the best to the crop which has no after effects on the quality and consumption of the product. Mixing organic manures at the time of soil preparation is very vital. Products of continental vegetables in the European and American markets are highly prized arrivals and also fetch very high prices in the market even 2 times than the traditional culture system.

Continental vegetables are adapted to low temperature or temperate status, require optimum moisture status in the soil and are very little dependent on irrigation. Irrigation thus required is very low under those conditions. However under Indian conditions dry spell status or prolonged dry spell or high temperatures during the growing period will lead to crop bolting and seeding of Lettuce, stunting growth of Broccoli, Celery, Parsley, no fruit formation in Brussels sprouts and more vegetative growth of Snow Peas and no fruiting. These are the causes related to the stresses with which a crop must have undergone before exhibiting the signs. Thus timely and proper moisture status and other management practices are the highly important critical inputs for these vegetables.

Vegetables in the field carry no value without post harvest handling system innovation for storage and marketing. It is here where the producer gets the highest value for his crop when proper technology of post harvest is inducted. Every crop is harvested with least damage and in the cool hours of the day i.e. morning or evening, subjected to pre-cooling, grading and storage under optimum conditions of temperature in the cold store. In the international market products need projection of customer appeal and attraction. Packaging is thus the ultimate step in presenting the product to the consumer, importer in the desired standards of demand.

Soil Analysis : Sustainable and productive culture of vegetable demands the diagnostic analysis of the soil by the farmer to know about the exact deficiencies and cater to the requirements with the addition of macro and micro nutrients in the soil being the demand of a particular crop. Soil analysis will indicate the pH value or the status of acidity which is vital in making available the plant nutrients to the crop. In case low pH value, soil requires addition of calcium. Soil analysis has an advantage of economic application of required nutrients and without increase in cultivation costs.

Nursery Raising : Continental vegetables belonging either to Perennial or Bi-ennial or annual crop category does require a special culture of raising seedlings prior to planting. Nursery raising is thus a process of growing young healthy plants with maximum care till these are capable of being transplanted in the field. This saves not only the seed costs for costlier seeds but provides utmost care and look after of the young seedlings. Seeds in the nursery are sown in well prepared well pulverized seed beds fully supplemented with organic manures. The seedlings are transplanted when these are either one month old or have developed 4-6 leaves, with a good root system. Plants are dug out without damage to the root system and transplanted in regular well prepared fields at pre-determined spacings.

Seed beds are prepared either by ploughing or digging with forks and pulverizing. Manures are then mixed and the bed is leveled. Nursery beds are either raised or sunken is the choice of the farmer. Seeds which take longer time for germination are protected with mulch so as to conserve moisture and restrict weed growth.

1. COLE CROPS

Origin : Cole crops formerly from the Cruciferae family, have been put in Family Brassicaceae with the new classification. In the Roman Empire Kale like ancestor was grown in the gardens. In Europe Cabbage gardens were very important food sources during the Middle Ages. Differences in morphology of Cole crops are the result of early selection for various edible parts. All Cole crops are Interfertile and many are self incompatible. These characteristics have made it easier to select for new types of Cole crops. Self incompatability makes Hybrid seed production much more economical. Cole hybrids are preferred because of their vigorous growth, uniformity of maturation and disease resistance. Over the last 3000 years, wild Cabbage (*Brassica oleracea* L. subsp. *Oleracea*; indigenous to the coastal areas of Western Europe and Great Britain has been developed into several distinct vegetables collectively known as Cole crops including Cabbage, Cauliflower, Broccoli, Brussles sprouts, Kale, Collards, Kohlrabi/ Knolkhol (Phillips and Rix). Related Brassica species have produced Chinese Cabbage, Turnip, Rutabaga and oil seed rape. *Brassica campestris* (rapa) is of Chinese origin and has got Pok-

Choy, Chinese Mustard, Celery Mustard, Chinese cabbage, Celery cabbage, Turnip broccoli in this group. *Brassica napus* group includes in itself Rutabaga – swollen hypocotyls, with yellow flesh, Turnip (Similar to Rutabaga, flesh white or yellow). Other species of the Family Brassicaceae used as vegetables are Radish, Chinese Radish, Horseradish.

BOTANY

Botanical Name : *Brassica oleracea*
Brassica campestris(rapa)
Brassica napus(rapa)

Cole Crops

Cauliflower
Broccoli
Cabbage
Brussels sprouts
Collards
Kale

CLASSIFICATION OF COLE CROPS

Brassica oleracea; subsp. var *capitata*; subgroup *alba*; Cabbage – heads formed by infloding of leaves, externally green, white internally.

Brassica oleracea; subsp. var *capitata*; subgroup *rubra*; Red cabbage – internally white and red

Brassica oleracea; subsp. var *capitata*; subgroup *sabauda*; Savoy cabbage – foliage highly wrinkled or savoyed, green.

Brassica oleracea; subsp. var. *botrytis*; subgroup *cauliflora*; Cauliflower – a tight flower, generally white, some purple.

Brassica oleracea; subsp. var. *botrytis*; subgroup *cymosa* (italicao; Broccoli – not as tight, more bud like individually, blue green in colour.

Brassica oleracea; subsp. var. *gemmifera*; Brussels sprouts – elongated stem, miniature heads(sprouts) in leaf axils.

Brassica oleracea; subsp. var *acephala*; subgroup *laciniata*; Kale, Collards – curly kale dwarf type, highly curled foliage.

Brassica oleracea; subsp. var *acephala*; subgroup *plana*; Smooth leafed Kale – rosette type, Collards in USA.

Brassica oleracea; subsp. var *acephala*; subgroup *millecapitata*; thousand head Kale – tall growing form, producing whorls of young shoots high above the soil.

B. oleracea; subsp. var. *acephala*; subgroup *palmifolia*; Tree Kale – tall (>2 m)

B. oleracea; subsp. var. *acephala*; subgroup *medullosa* marrow stem Kale – long thickened stems, occasionally curled.

B. oleracea; subsp. var. *gongylodes*; subgroup *caulo-rapa*; Kohlrabi – thickened stems 2-7 cm above soil, forming an above ground tuber like structure.

Brassica campestris; subsp. var. *chinensis*; Pok-choy – Chinese Mustard, celery Mustard, resembles cos. Letuce.

Brassica campestris; subsp.var. *pekinensis*; Pe-tsai – Chinese cabbage, celery cabbage, resembles Swiss Chard.

Brassica campestris; subsp. var. *ruvo*; Broccoli raab – rapa, rapine, turnip broccoli, grown for greens, un-opened flowers and stems.

Brassica napus; subsp. var. *rapifera*; Turnip – similar to Rutabaga, flesh white or yellow

Brassica campestris; subsp. var *napobrassica*; Rutabaga – swollen hypocotyls, flesh yellow.

VARIETIES

- **Broccoli**: Bonanza hybrid, Green Goliath, Packman, Spartan, Zeus hybrid
- **Red Cabbage** : Red acre, Red Drum head
- **Brussels sprouts**: Jade cross hybrid, Long island improved, Oliver hybrid
- **China cabbage**: Two seasons, Dynasty, Mitchhili, Wongbok

COLE CROPS

- **Cauliflower :** Katki, Early Kwari, Agheni, Poosi, Maghi, PSBK-1, Snow Ball-16, Holland Special, INDAM Early, Pusa Early Synthetic, Pusa Deepali, Pant Cauliflower 3, Pant Cauliflower 2, Pusa Kartiki, 234 S, 75-2°C,

- **Cabbage :** Hybrid CB-296, IAHS-3, IAHS-2 (Krishna), Indo-American Express, Ruchi (Red Cabbage). OPEN POLLINATED Golden Acre, Krishna, CB-296, 1299, 30297, 30997, INDO-AMERICAN Express, Chaubatia Early, Pusa Drumhead, Late Large Drumhead, Pusa Mukta, Copenhagen Market, RED CABBAGE–Red Drum Head, Red Rock
- **Knolkhol :** White Vienna, Purple Vienna
- **Broccoli :** Packman, Pirette, Green Head, Italian Green, DPGB 1
- **Brussels Sprouts:** Early Dwarf, Dwarf Gem, Danish Prize, Express, Sweden, Catskill, Long Island, Jade Cross

Maturity Period : Maturity period in case of Broccoli is 55-77 days but for the Pink Broccoli which takes 4-5 months. Red cabbage matures in 55-78 days after transplanting and Brussels sprout takes 85-100 days. China cabbage is again a 65-70 day crop in the field.

Seed Rate : For the Cole crop group vegetables, the seed rate is 125-150 gm per acre or 300-400 gm per ha.

Soil Preparation : Cole crops do better in medium to heavy soils rich in organic matter. Soils with a pH lower than 6.5 be supplemented with calcium applications. Field is well prepared prior to transplanting of seedlings and the seedlings are either transplanted on the raised beds or in plain. Manures and fertilizers are mixed well into the soil.

Spacing : Plants in this group are generally recommended to be spaced at 2 x 2 ft. or 60 cm both row to row and plant to plant.

Manures and Fertilizers : Application of organic manures is good for high productivity. Recommended schedule of the combination for the group is —

	Constituent Requirement kg / acre		
	N	P	K
	30	40	20
Neem cake	2 qtl / acre		
Animal Meal 7:10:0	2 qtl / acre		
DAP	34 kg / acre		
Muriate of Potash	35 kg / acre		

20 kg of Nitrogen in the form of Urea is provided as top dressing per acre in two split doses after 25 and 45 days of transplanting.

Irrigation : In the hill regions where irrigation is not available, mulching is done to conserve moisture and keep optimum moisture level for the crop and to save it from irrigational or moisture stress. In the plains where temperatures at one time will tend to be higher, irrigation is required.

Interculture : For a non competitive crop stand with weeds crop must be kept free from weeds by frequent intercultural operations and weeding.

DISEASES AND PESTS

- **Damping Off :** Causes softening of the seedlings stem, plants lie down after attack and die. To ward off from this disease the seeds are treated prior to sowing with CAPTAN or BAVISTINE and nursery beds are sterilized with Formaline solution 3-4 days prior to nursery sowing. Beds are checked for formaline odour in the soil. When temperatures are high apply Bavistine @ 1-2 gm dissolved in 1 liter of water every 4-5 days.
- **Cabbage Root Maggot and Aphids :** These are the two major pests of this crop group. At the time of transplanting dip the roots of the seedlings in DIAZINON. For control of Aphids Tobacco decoction spray is highly useful and organic too.

Harvesting : Harvest central heads of Broccoli when crop harvest maturity is there. For Red cabbage harvest highly compact and fully mature heads. In the case of Brussels sprouts tiny cabbages at the bottom of the stem are harvested first. Top sprouts are let to improve in size and compactness.

2. LEEK

Leek *Allium porrum*

Leek is a member of the onion family. It is in much demand because of its culinary value and subtle flavour and substitute for onion. Highly blanched stems of Leek, cylindrical in nature are much prized in quality acceptance in the market. 6-9 inch blanching of the Leek bulbs is best quality production. Leeks make superb soups and are indispensable in French cooking.

Varieties : Prize Taker, Giant Mussel burgh, Titan, Broad London.

Days to Maturity : 140 -150 days

Seed Rate : 3 ½ - 4 kg per care

Nursery Raising: In the high hill zones of Himachal Pradesh, Uttaranchal, Jammu and Kashmir, Arunachal Pradesh seeds are sown in the nursery in the months of January end to March. In the plains of India seeds are sown as per Onion culture in the months of October – December.

Spacing : Plants are spaced 4 inch's with a distance of 2 ft. from row to row.

Soil Preparation : Prepare the soil either by tractorization for large plantings or by digging for small area cultivation. Mix all the manures and fertilizers as per recommended doses. Plank and level the soil and planting bed.

MANURES AND FERTILIZERS

	Constituents Required Kg / Per Acre		
	N	P	K
	60	25	40
Basal Dose Per Acre			
Neem Cake 1 qtl	5.2	1	0
Agromeal 7:10:0	7	10	0
DAP 35 kg / acre	6	14	0
CAN 1 qtl / acre	26	0	0
Muriate of Potash	0	0	40

Urea 46%, 50 kg is applied in 4 split doses of 12½ kg per care each 50 days after transplanting at an interval of 20 days.

Interculture : High earthing up is done to blanch the bulbous stem of Leeks for quality maintenance. This also keeps the soil free from weeds.

Irrigation : Keep the soil at the optimum moisture level by applying irrigation when the soil seems getting dry.

Harvesting : Leek is winter hardy herbal vegetable as well as withstands high summer temperatures. Leek does not like water stagnation or prolonged high water status in the soil. Bulbs start rotting during rainy season in the plains of India. Leek varieties which tiller profusely or bring out lateral shoots from the base are the best source of future propagation. This quality of the variety is highly important in shortening the cultural time of the next crop thereby bring the time to 70 – 80 days for crop maturity. For quality harvest, Leek should be harvested prior to tillering in these varieties. Leeks are mature to be harvested when the leaves are broad enough. Each plant weighs 80 – 120 gm.

Yield : Yield of 50 – 105 qtl / acre

Seed Production : Leek is a Biennial crop and the seeds are obtained in the second year of crop culture from the Leek bulbs of the earlier crop, which flower and produce seeds like Onions.

3. CELERY

Celery *Apium graveolens*

Celery is a member of the family Umbelliferae. Leaves and the stalk bear a typical flavour highly prized in soups, dressings and garnish. Leaf stalks are highly juicy. Highly blanched celery is preferred in the market. There are two types of Celery i.e. Non-Blanching or Trench Type and Self Blanching. Non-blanching types are the same as Indian Ajmod. European celery cultivation methods is much different than the Indian. In the hills care is taken to avoid frost damage to the plants. Harvest maturity depends on the heart development, with thick dark green long stalks.

Varieties : Tall Utah, Ford Hook, Golden Self Blanching, Wrights, Giant White

Days to Maturity : 105 – 130

Nursery Raising : Raise nursery for hill region sowing from the 15th January onwards and continue till March-April. In the plains of India start from the September month and continue till December. Nursery time takes 35-40 days for seedlings worth transplanting.

Seed Rate : 100 gm per acre or 250 gm per hectare is the seed rate.

Plant Spacing : Plants during transplanting are spaced at 6-9 inch from plant to plant and rows are spaced 1½ - 2 ft. apart. The spacing depends on the habit of the variety.

MANURES AND FERTILIZERS

	Constituents kg per acre		
	N	P	K
	60	20	-
BASAL DOSE PER ACRE	40	20	-

Agromeal 7:10:0 2 qtl / acre

Urea 46 % 66 kg / acre

Top dress celery crop with 3 split doses of Nitrogen from Urea @ 14 kg/acre per dose repeating after every 15 days of first dose.

Soil Preparation : Prepare the soil thoroughly making it fully pulverized by giving 2-3 deep ploughings and mixing all the basal dressing manures. Plank and level the beds. Celery requires highly porous soil for its good growth.

Irrigation : Keep the soil moist with optimum moisture status. In the high hill zone where irrigation is not possible, make use of poly sheet mulch's or grass mulch, which will conserve moisture loss and also discourage weed growth. In the plains irrigation frequency be kept at 5-6 day interval period during the winters and be decreased to 3 days when temperature start rising.

Interculture : Fields should be kept free from weeds and always porous for good air circulation for the roots and its growth. In the case of varieties to be blanched, earthing up is desired of the stalk.

Harvesting : Celery is harvested at the desired maturity status when the leaves are fully grown, green and the stalk well developed and blanched.

4. PARSLEY

Parsley *Petroselinum crispum*

Parsley is a green herb for use in decoration, flavour enhancer, garnish and seasoning of dish's. There are three main types of Parsley for crop culture. These are Beaked or French, Italian or Plain, Curled or Moss curled. Varieties are also named as per their character. Chervil (*Anthriscus cerefolium*) is also a member of this group of herbs. In the Landscape designing and also as a companion crop Parsley is used as an edging crop in the gardens. Classification of the parsley as per their origin is:

- *Petroselinum filicinum* or neopolitanum is Italian Parsley.
- *Petroselinum crispum* is called Curled or Moss curled Parsley.

In India market is developed for Moss curled varieties of Parsley only. Chervil or Italian Parsley possesses strong flavour and are superior in cooking. Curled Parsley are crisp.

Varieties : Extra curled dwarf, Plain Italian, Single, Moss curled

Days to Maturity : 70 – 90

Seed Rate : 400 – 500 gm per acre

Nursery Raising : Parsley can be sown directly into the prepared fields, but nursery raising is safer to get healthy plants and save on seed costs as the seed rate then gets increased to 2 – 3 times than the nursery seed rate. In the direct sowing, plants are thinned to provide proper spacing for each plant to develop normally. Plants are developed from nursery and transplanted at the desired spacing. Seedlings develop within 30-35 days for transplanting in the field.

Spacing : Distance between plants is kept in the range of 3-6 inch and the rows are spaced 9 – 18 inch from row to row.

Soil Preparation : Soil preparation is the same as for celery.

Manures and Fertilizers : Animal manure or Agromeal 7 : 10 : 0 is applied at the time of soil preparation @ 2 qtl. Per care. Two split doses of Nitrogen application are done @ 8 kg Urea 46 % after 40-60 days of transplanting. This provides a very good vegetative growth to the plant and yield gets increased.

Hoeing and Weeding : Crop takes nearly 3 months for maturity, thus it must be well protected from weeds and the soil be kept well porous for proper development of the plant.

Harvesting : Parsley is meant only for harvesting of foliage for the market, as such well grown up and developed leaves or fully curled leaves are harvested every 3rd or 4th day. Basal leaves are the first to be harvested.

Seed Production : Parsley is a cross pollinated crop, as such due precautions in this are required to be undertaken. This requires a distancing of atleast 1000 mtrs from the other varieties. Harvest the Umbels prior to seed ripening, since the Umbels has the tendency to shed after ripening. Clean and store the seed properly. Seed production expected is 250-300 kg per acre.

SALADS

Vegetables consumed in un-cooked form are known as Salads. Lettuce and Endive / Escarole are two such crops in this category, which are prized for their usage and culture. Another name which has been added to this group is Belgian Endive Chicoons.

5. LETTUCE

Lettuce have its origin in Egypt. Evidence from Egyptian Tomb paintings suggest that Lettuce was cultivated before 4500 B.C. It was prickly lettuce which was originally cultivated for forage and oil. Prickly Lettuce (*Lactuca serriola)* is extremely bitter. This bitterness is associated with the production of latex, the milky juice. This milky juice is sleep inducing. Romans developed non heading broad leaved, non spiny types that were resistant to early seed stalk formation or bolting, with decreased latex. Romans also blanched Lettuce to make them less bitter. Lettuce the milky exudation of the leaves is very healthy and medicinal in its properties and acts as a tonic for the stomach. In Ayurveda consumption of Lettuce has been advised for toning up the digestive system.

BOTANY

Family : Compositae
Lettuce: *Lactuca sativa*

Lettuces can be divided into four sub groups —

- **Loose Head (Leaf) :** Easy to grow, thrives best even under the poorest of the poor soil conditions. Plants are fast maturing, produce large round thick bunches of attractive leaves that are crisp, tender and mild. Intensely coloured leaves are more nutritious because more Carotenoids and Flavonoids. Varieties mature in 45-50 days. Varieties are Green Ice, Royal Oak Leaf, Tango, Red Sails, Red Salad Bowl, Grand rapids, Ruby.

- **Cos or Romaine :** Upright growing, tightly folded heads with light green elongated leaves developing into large loaf shaped heads. Slower to bolt than other Lettuces. Useful as a warm weather crop. Varieties in this group are not many, but Paris White Cos is the one known, which matures in 80 – 85 days.
- **Butter Head :** Hearts are tightly folded and blanch to a golden or buttery tallow, are tender with tasty leaves, grow rapidly. Soft leaves with smooth texture. Varieties bred for both outdoor summer conditions and Green House winters in Europe and USA. Summer Butter Head varieties are larger and firmer than the Green house varieties. Varieties are Bibb and Butter Crunch, Dark Green Boston. These mature in 75 days from the day of transplanting.
- **Crisp Head** : Forms large, heavy, tightly well folded heads of crisp, brittle texture. Crisp Lettuce leaves are prominently veined, wrapper leaves green, inner leaves whitish – yellow. These are predominantly outdoor types. These make very large round heads when grown at recommended spacings. This group blanches silvery white or creamy yellow in the center. Variety withstands more heat than other sub-groups. Varieties in this sub-group are Great Lakes, Burpee's Iceberg and Minetto which mature within 80-90 days from the transplanting day.

Days to Maturity : Maturity period differs from each group from 45-90 days taking the least for Loose Head and maximum of 90 days for the Iceberg.

Seed Rate : 80-100 gm per acre

Nursery Raising: Lettuces can be sown directly or seedlings raised through nursery. Seedlings get ready within a period of 30 days for transplanting in the field. Seedlings are ready when 3-4 leaf development is observed. Over grown seed-lings in the nursery be avoided as this takes to bolting and smaller head development.

Spacing : Lettuce plants are spaced at 1 x 1 ft or 1 ½ x 1 ft distance. The distances are purely dependent on the type of the sub-group and the variety.

Manures and Fertilization : In Lettuce culture desired growth is for the green heads, as such only application of Nitrogen is recommended.

Constituents per acre		
N	P	K
30	-	-

BASAL REQUIREMENTS :

Neem de-oiled cake 6 qtl / acre

Or Animal meal / Agromeal 7:10:0

Apply Neem de-oiled cake or 4½ qtl of Agromeal at the time of soil preparation. No top dressing is done. When plantation of Lettuce is in Garden type culture, liquid application of Animal meal is recommended @5 kg in 20 liter water soaked over a period of 8-10 hours and then further diluted to 100 liter. Application of liquid manure @ 100-200 ml per plant gives best results.

Sowing Time : Lettuce cultivation in the hilly regions starts from the month of March and continues till July. In the plains the culture is undertaken during the winter months as a Rabi crop and it starts from the month of October and continues till January. During the Rabi season a farmer can specialize to grow atleast 3-4 crops of Lettuce from his fields, by rotating the culture after every 45-50 days for open leaf and 2 crops for other subgroups.

Soil Preparation : Soil is prepared by ploughing or digging of the beds and whole basal dose of manures is mixed in the soil.

Transplanting : Bold and healthy seedlings possessing 4 leaves are taken out in the morning and firmed in the beds and irrigated immediately. Transplanting is recommended to be undertaken during evenings, so that the seedlings get good low temperature for the root setting in the soil.

Irrigation : During winters in the plains the irrigation is done after 5-6 days, but when the temperature start rising after March, frequency of irrigation is kept at 3 days interval.

Interculture : During the first 20 days of transplanting weeding should be given proper care. With the spread of the heads, these take care of themselves suppressing the weeds underneath. Mulching is recommended to reduce cultivation costs.

Harvesting : Heading Lettuces are harvested when the heads are tight enough. Loose Leaf lettuce are harvested when the head dia is 9-12 inch and quite compact and full of leaves. Cos or Romaine varieties are ready for harvest when good number of leaves have developed and leaf size is big enough.

6. ENDIVE

Endive and Escarole *Cichorum endivia*

Witloof Chicory or Belgian Endive *Cichorium intybus* var. *folosum*

Vernacular name of Endive in India is Kasni by which it is popularly known. This belongs to the Compositae Family and is a well known Gourmet Salad herb of the Mediterranean region being native to the area. This is a cool season or winter crop for the plains of India and summer crop for the High hill zones of the Northern India. It is cultivated in Northern India and leaves are eaten raw. As a continental vegetable this has been reckoned with Lettuces for its culture and usage. Main varieties developed are Green curled Endive or Giant Fringed Oyster, Escarole or Broad Leaved Batavian. Both the varieties take 90 days for full maturity in the field. The heads of Endive are blanched from the center to provide it a much better look and acceptance in the market. Belgian Endive/Witloof Chicory, is *Cichorium intybus* var *folosum* and this is chance discovery by a Belgian farmer in the year 1850. Harvested roots of Chicory were defoliated and kept in a dark store house for 15-20 days. These roots developed 3-4 inch chicoons during this period, which were very tasty, sweet. Right since then this formed a special cultural item of the Chicory roots as Witloof Chicory or Endive. The production of Chicoons is a process of forced culture. Witloof chicory be not confused with the chicory whose dried roots are used as coffee.

Broad crumpled leaf types of Endive is the specie endivia and also known as Escarole. This has very many different varieties developed for cultivation under different conditions. Broad Batavia Full hearted, Grosse Bouclee (Slow bolting), Salanca (cold tolerant) are some of the varieties. In India no major research or development work for this has been done. As such Indian grower of these vegetables has always to depend for his seed requirement from European or American Seed Companies.

CULTURE

Soil : Endive or Escarole is grown on a very wide range of soils. It requires loose fertile loams with very high organic content and good water holding capacity and much better drainage. Endive loves a pH of 6.5 and above.

Sowing Time : For early crop sowing of seeds in the plains is done under cold conditions of Green House in the month of September and, seedlings are then transplanted during the beginning of October. Sowing of nursery is undertaken in the plains of India till the 15th January to get late crop for the market. In the hills of India, it is sown during March and continued till June.

Propagation: It is propagated by seeds only either by directly sowing the seeds or raising the seedlings in nursery beds. Optimum nursery temperature should be 18.25°C for good nursery crop. One gm of Endive or Escarole seeds have nearly 750-850 seeds.

Soil Preparation : Soil is prepared by giving two deep ploughings coupled with harrowing and planking. Basal dressing of FYM is given with the preparation of soil @ 10-20 tons per acre.

Spacing and Density : Plants are spaced at 15-18 inches apart in the rows and distance from plant to plant is kept at 10-12 inch, thereby giving a density of 29000-35000 plants per acre.

Manuring and Fertilization : Fertilizer recommendation for Indian conditions is not available because of no such research work, however European recommendations are for the supply of 45 : 45 : 25 kg NPK nutrients / acre. Nitrogen is applied in 2-3 split doses with Phosphorous and Potash as basal doses.

Irrigation : It requires a constant moist status of the soil for good head growth and addition to weight. During winters irrigations are given after every 8-10 days and starting March frequency of irrigation increases with the rise in temperature.

Bolting : Bolting of the Endive or Escarole heads can be triggered by plant stress due to flooding or hot dry weather, very low or high temperatures, long days. These conditions should not be let to built in during the crop culture to obtain good crop yield.

Yield : Average yield for Endive or Escarole has been recorded in the range of 6.5-9.5 tons per acre with the above density plantations.

Harvesting and Storage : Endive and Escarole heads after harvest are culled down immediately. Heads are stored at 0°C with 95% RH level in the store. Under this temperature regime Endive can not be stored for more than 2-3 weeks. Other processes of cooling which have been found satisfactory are Vacuum or hydro-cooling.

Raising of Chicoons : Green foliage from Witloof Chicory Roots is cut and the roots 4-5 in number are placed in sand media in 12 inch earthen pots in the months from January-March. These pots are then watered copiously. Another earthen pot is placed upside down and these are covered with a black poly bag to provide dark condition the basic requirement along with humid moisture conditions. After 10-15 days chicoons of trade and commerce will come out from the roots. These are severed and sold in the market as Salad crop.

7. ARTICHOKE

Introduction : Artichokes of trade and commercial culture are Globe Artichoke, Jerusalem Artichoke and Chinese Artichoke. These are very different to each other in their botany, culture and the edible parts and need not to be confused with each other. Globe Artichoke is the developed flower and Jerusalem Artichoke is the root or underground rhizome of the plant. As a vegetable this seems to have been neglected only because of the plant structure and has never tried to taste it as a vegetable. Artichokes whether flower or under ground rhizomes are delectable vegetables which are a very good source of dietary fiber, vitamin C, folate (folic acid) and very many minerals. Each cone shaped un-opened flower bulb of Globe Artichoke *(Cynara scolymus)* is overlapped by tough and in-edible scales but is fleshy and tender at the base and inside. This inside fleshy portion is called the heart. Globe Artichoke is a warm climate vegetable and grown all over the world. Artichokes are harvested through out the year with peak supplies from the spring till May and again continuing to a lesser extent up to October. Un-opened flower buds are unto 4 inches in diameter and are either spherical or oval.

Globe Artichoke: Edible portion in this is the immature globe type flower bud.

Jerusalem Artichoke : Edible part in this is the tuber and resembles very much like Potato.

Chinese Artichoke : Edible part in this is also the tuber.

BOTANY

English/Local Name	Botanical Name	Family
Globe Artichoke	*Cyanara scolymus*	Compositae
Jerusalem Artichoke	*Helianthus tuberosum*	Compositae
Chinese Artichoke	*Stachys affinis*	Labiatae
	S. seiboldi	Labiatae

Jerusalem Artichoke : also known as Sunchoke. Fresh tuber tastes like a water Chestnut and is used in Salads. Tubers can also be cooked like Potatoes. Edible portion which is the tuber or swollen end of underground stem resembles in very many respects with Potato. However, unlike most starchy vegetables (Potato), the principal storage ingredient Carbohydrate in Sunchokes or Jerusalem Artichoke is Inulin rather than starch. When consumed, Inulin gets converted to Fructose rather than Glucose, which is tolerated by diabetics. Also Jerusalem Artichoke is increasingly popular as a slimming food and takes 90-120 days for crop culture.

Chinese Artichoke: Both *Stachys affinis* and *S. seiboldi* are tuberous crops and are eaten either raw or as a cooked vegetable. *S. seiblodi* has long slender tubers. Both these are a favourite vegetables in China and none where else in the world.

Varieties

- Mammoth French White
- Mammoth French White Improved
- Stampede White, Early yielding 90 day crop.
- Fuseau White, Long, straight, knob free tubers
- Red Fuseau Maroon red skinned, Early yielding, elongated, smooth
- Dwarf Sunray English variety, tender, crisp, no peeling required
- Brazilean White
- Brazilean Red

Climate : This crop is a cool climate loving and is better adapted with the onset of winters in Indian climate. North Indian states of Himachal, Punjab, Haryana, Jammu and Kashmir, Uttaranchal and Uttar Pradesh, are some of the best states. In Uttar Pradesh the Gangetic and Yamuna belts are the ones which with their sandy belts suit for the culture of Jerusalem Artichoke during winters.

Jerusalem Artichoke

Soils : Well fertile sandy loams or well drained river belt beds are the best for its culture. However where Potato and Maize are cultivated are the best soils for Jerusalem Artichoke culture too.

Planting : Tubers no less than 50-60 gms weight be planted. Tubers are planted 6 inches deep into the soil with a spacing of 15-24 inches from plant to plant and 3-3 1/2 ft. between rows. Plant densities vary with the plant to plant and row to row distance. The ones which are suited are as under —

PLANT DESNITY PER ACRE

Distance in Ft.				
Plant to Plant	Row to Row	No. of Tubers	Seed Qty. in Kgs.	Expected yield in MTs
1	3	14250	726	14-21
	3.5	12245	622	12-18
	4	10890	544	10-16
1.5	3	9680	484	9-14
	3.5	8297	414	8-12
	4	7260	363	7-11
2	3	7260	363	7-11
	3.5	6222	311	6-10
	4	5445	272	6-8

Sowing Time : The crop matures in 90-120 days time and loves cool condition. Sowing time is thus very critical when there is no heat at the time of harvest maturity. This time thus is the same as that of the Potatoes. October-December are the best periods for the successful culture of Jerusalem Artichoke.

Cultural Management : Grasses and the weeds are the main problems immediately after sowing and taking a good crop stand. Thus this requires very frequent weedings for the removal of grasses and weeds from the fields. No sooner the crop attains good growth, it has the power to suppress and control the weeds automatically then due to shading of the soil. Apply irrigations frequently for good growth.

Fertilization : Jerusalem Artichoke unlike all other tuber crops loves high nutrient applications of NPK. Applications of these be made strictly according to the soil analysis. The general recommendations are:-

- N 60 – 90 kgs/acre
- P 35 – 65 kgs/acre
- K 35 – 45 kgs/acre
- S 5 - 10 kgs/acre

Harvesting : can be done both manually and mechanically as in Potatoes. Large woody foliage is removed before harvesting the tubers. Jerusalem Artichoke is not a commercial crop in India as such only manual harvesting is done. Furrows are opened with spades or plough and hand picking of tubers is done.

Handling and Storage : Skin of Jerusalem Artichoke is very thin, as such care should be taken while handling to avoid skinning, cuts and bruises. Avoid moisture while handling and storage operations by storing tubers at 0°C or 32°F and RH of 90–95%.

Yield : depends very much on plant density/ population, length of season, cultural management practices. Yield of tubers averaged 29 tons/acre in a 3 year study in South Western Oregon. Yield also depends very much on top removal, row spacing.

Usage : Source of raw product for Fructose sugar and Alchohol production. It is also a very good Live stock fodder and feed.

Seed Availability : Though seeds of Jerusalem Artichoke are available in India only in selected Seed houses, but varieties are un-named. However for a quality production, Germplasm and seed material can be obtained from foreign countries.

- Johnnys Selected Seeds, ALBION, MA 206-437-9294
- Garden City Seed, VICTOR, MT 406-961-4873
- Southern Exposure Seed Exchange, Earlysville, VA
- Guernys Yankton, SD 605-665-1930
- Thompson and Morgan Suffolk, England

8. GLOBE ARTICHOKE

Introduction: Artichokes of trade and commercial culture are Globe Artichoke, Jerusalem Artichoke and Chinese Artichoke. These are very different to each other in their botany, culture and the edible parts and need not to be confused with each other.

Globe Artichoke : Edible portion in this is the immature globe type flower bud.

Jerusalem Artichoke : Edible part in this is the tuber and resembles very much like Potato.

Chinese Artichoke : Edible part in this is also the tuber.

BOTANY

English/Local Name	Botanical Name	Family
Globe Artichoke	*Cyanara scolymus*	Compositae
Jerusalem Artichoke	*Helianthus tuberosum*	Compositae
Chinese Artichoke	*Stachys affinis*	Labiatae
	S. seiboldi	Labiatae

Nutritional Value : Globe Artichokes is a highly nutritious vegetable having not only the dietary fibers but also vital ingredients, minerals contained in it. WholehealthMD.com has reported for its nutritional values of one cup of Artichoke flowers containing 76 calories of energy, 0.2 gm total fat (0.1 gm saturated fat, no monounsaturated fats, 0.1 gm polyunsaturated fat), 8.8 gm dietary fiber, 5 gm protein, 17 gm carbohydrates, 152 mg sodium, 19 mg Vitamin C, 110 mcg folate, 0.4 mg copper, 97 mg magnesium, 0.4 mg manganese, 146 mg phosphorous and 599 mg potassium.

Globe Artichoke : True Artichoke, member of this family known to the trade and commerce as Globe Artichoke. Globe Artichoke is the developed flower and Jerusalem Artichoke is the root or underground rhizome of the plant. As a vegetable this seems to have been neglected only because of the plant structure and has never tried to taste it as a vegetable. Artichokes whether flower or under ground rhizomes are delectable vegetables which are a very good source of dietary fiber, vitamin C, folate (folic acid) and very many minerals. Each cone shaped un-opened flower bulb of Globe Artichoke (*Cyanara scolymus*) is overlapped by tough and in-edible scales but is fleshy and tender at the base and inside. This inside fleshy portion is called the heart. Globe Artichoke is a warm climate vegetable and grown all over the world. Articholkes are harvested though out the year with peak supplies from the spring till May and again continuing to a lesser extent up to October. Un-opened flower bulbs are unto 4 inches in diameter and are either spherical or oval.

Varieties

- Green Globe Improved Thornless, green purple tinting, Globe shaped heads, with a very heavy bearing habit.
- Imperial Star
- Emerald
- Grand
- Talpiot
- Northern Star

Propagation : It is propagated from seed, crowns and side shoot divisions. Seed propagated varieties are not true to type because of variation. Best process of propagation is resorted however through seeds because of not much material available in the form of crowns or side shoots. In the Western countries growing period reduction or fruiting/flowering period shortened by vernalization of seeds. While growing Artichoke from seeds, it is grown as annual where it is not likely to survive winter due to freezing conditions. Seed germinated plants are grown and set out in fields with a spacing of 2 x 3 ft from plant to plant and 3x4 ft. from row to row. When this crop is to be made permanent, then alternate rows are of plants are dug out and accommodated elsewhere in other fields.

Vernalization of Globe Artichoke seeds is not a recommended practice, but it is 2-4 weeks exposure of seeds at about 40°F or 5°C exposure for a set time of period which can be anywhere between 4-7 days. Opinion of Connecticut Researchers recommend pre-soaking of seed for 48 hours to soften the seed coat and then holding these moist seeds for 4 weeks at temperature between 35-40°F or 1-5°C in Sphagnum Peat or Cocopeat.

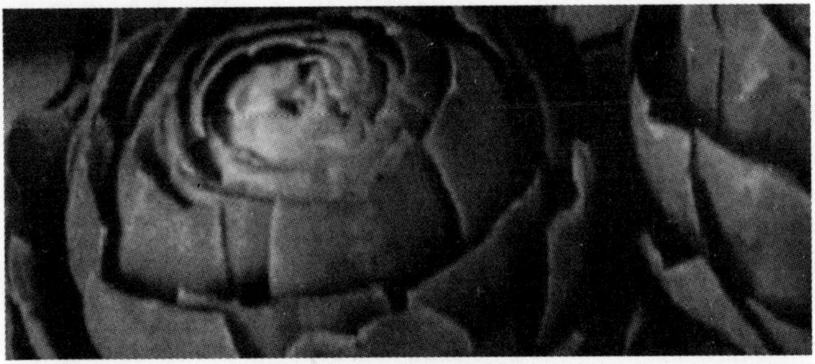

Globe Artichoke

Plantation Density: Plant population per acre or density of plants is different both for annual and permanent plantings dependent on plant to plant and row to row spacing. In the case of permanent culture density is anywhere between 1000 – 2000 plants per acre. In the case of annual culture it can be 4000-7000 plants per acre. 28-30 seeds of Globe Artichoke weigh 1 gm. Seed rate thus for permanent and annual crops is around 40-90 gms and 130-250 gms per acre.

PLANT DENSITY PER ACRE
DISTANCE IN FT.

Type of crop	Plant to Plant	Row to Row	No. of Plants per acre
Permanent	3	6	2420
		7	2074
		8	1815
		9	1613
		10	1452
	4	6	1815
		7	1555
		8	1361
		9	1210

		10	1089
Annual	2	3	7260
		4	5445
	3	3	4840
		4	3630

Fertilization : Well prepared fields are manured very heavily with animal or organic manures at the rate of 20-30 tons per acre. NPK are recommended to be applied after the soil analysis and range from 45-90 kgs of each constituent/acre. Nitrogen application be applied in split doses i.e. 1/4th before the planting of the crop and the rest after 6-8 weeks from transplanting. In permanent Globe Artichoke culture application of Nitrogen in the second year is reduced to 25-45 kgs/acre. Sulphur is applied @ 6-10 kgs/acre and is broadcasted and worked in the fields.

Irrigation: Globe Artichoke is a deep rooted crop, requires 15 inches of irrigation during its growing period. Irrigation stress may lead to physiological disorder of Black Tip.

Use of Growth Regulators : To enhance earliness and uniform flowering of the crop, use of Gibberellic acid is recommended at 20 ppm making 2-3 sprays at 2 week interval in water 100-120 liters per acre. Foliar application of GA is recommended to commence after 5-7 weeks of transplanting when the plants are 1-1½ ft high. Due precaution is also to be taken for any mis-application due to higher rates of GA than the recommended ones and also spraying at high temperatures during the day. This may effect to reduce plant vigour and cause elongation of buds.

Harvesting: Fully grown immature, un-opened flower buds, composed of cone of bracts with good size and compactness are selected for harvesting. Other quality indices for marketable buds is a typical green colour, stems cut 1-1½ inch below the flower base. Harvested buds are then placed in crates to avoid harvesting and handling damage.

Quality of the Flower : Whatever is the size or shape of the Artichoke flower bulb it should be compact, heavy for its size, with scales or leaves on it that are tightly closed, fleshy, firm and thick. This is the harvest maturity stage and the quality of the flower.

Flowers looking dry or with open scales or scales having started spreading from the cone these are not marketable.

Post Harvest Processing : To maintain quality and increased shelf life through storage conditions harvested flower buds are immediately put to pre-cooling at temperature 40°F or 5°C through the process of Hydro-cooling or Forced Air cooling or Package-icing. This is done to reduce respiration rate of live buds which is 8-22 ml CO_2/kg. hr at 0°C and increases with the rise in temperature reducing its quality. Storage potential of Globe Artichoke is generally less than 21 days as visual and sensory quality deteriorates rapidly.

Packing : In USA Globe Artichoke is commonly packaged in cartons and Boxes 7 inches deep to provide a standard weight to the consignment with specific number of buds in the carton or box. Various grades on the diameter of the bud which is reckoned from the center of the bud are as under —

Diameter of Bud in inches	No. of flower buds Per carton/Box
4.5 and above	18
4– 4.5	24
3.5 – 4	36
3 - 3.5	48
2.75 – 3	60
1 - 2.75	Small loose

Market Demand: Artichokes are generally demanded garden fresh in the market. However it is also available in other forms as frozen and in Cans and Jars. Canned Artichokes come in brined form and are ready to eat stage. In the Jars marinated flower hearts are seasoned with olive oil or oil and vinegar mixture as a ready to use item. Frozen hearts have no added ingredients but are ready to be cooked after thawing.

Storage : Artichokes is a highly perishable commodity but are hardy to withstand storage. After washing and rinsing flower hearts place these in plastic packs, seal and create Modified Air Packaging conditions and place in cold chambers at 0°C or 32°F with a RH of 90%. Such storage has the capability of increasing the shelf life of the commodity up to 2-3 weeks.

Pest and Disease Control : Field selection, proper management, rotations, spacing of plants, sanitation can reduce the pest and disease problems to a greater extent.

9. ASPARAGUS

Asparagus has been found wild in many places and origin is not clear (Boswell). However the origin is placed to Eastern Mediterranean, Asia Minor (Boswell, Sturtvent, Vavilov). It is under cultivation for the past 2000 years. Asparagus was a prized food and recipe with Greeks and Romans (Thompson and Kelly) and it was valued for its medicinal properties contained in all parts of the plant. Properties of Asparagus were well recognised as medicine for almost anything from the prevention of Bee stings to Heart trouble, dropsy and tooth ache. Romans were well conversant of Asparagus culture before Greeks, as early as 200 B.C. (Sturtvent). Europeans and Britons have been eating Asparagus for as long as records exist. In America it was introduced by Voyagers (Boswell). Wild Asparagus is sweet, more pleasant to eat (Pomponius, Palladius). Word Asparagus originates from Greek word ASPARAGOS or ASPHARAGOS meaning sprout or shoot. In the 16th century England the word SPERAGE was used to describe the plant (Evelyn), where in his diary he mentions "SPERAGE, delicious, eaten raw with oil and vinegar". It is Asperge in Dutch and French, Spargel in German, Esparago in Spanish.

CLIMATE REQUIREMENTS

Asparagus though a plant of the temperate region but has the capability to adapt very well to different climatic conditions and can withstand both hot and cold extremes. Areas with monthly average temperature of 60-75°F and a winter dormant period produce good crop. Excess water causes root asphyxiation. Branching is an undesirable characteristic which is the phenomenon of high temperature extremes. Strong drying winds retard growth on one side of the spear causing crooked spears as an un-marketable lot. Under Indian conditions where every type of climate from temperate to tropical is available, cultivation of Asparagus stands the highest success rates in its culture. In the tropical culture the major requirement will be that of providing a dormant period in the culture which can be forced in the crop with the **Taiwanese** system i.e by forced cutting back of the plants during December and allowing a comfortable period for the crown and fleshy roots to develop for the next crop. After the stoppage of harvesting during June or July the plant gets ample time for the manufacture and storage of food for the next season.

BOTANY

Family : Liliaceae

Bot. Name : *Asparagus officinalis* var. *altilis* Asparagus is indigenous to temperate Europe and Asia where it has been in cultivation for the past 2000 years. It has been introduced into very many parts of the world from there. For Greeks and Romans this was very important medicinal crop. Asparagus belongs to Lily family and includes plants such as Onion, Garlic, Leeks, Shallots, Lilies, Tulips, Gladiolus, Hyacinths. Genus Asparagus has nearly 150 species both ornamental and edible. The garden or cultured specie *A. officinalis* var. *altilis* has also been described by Bailey.

Asparagus Jersey Knight Male

This is a hardy perennial, grown for its edible spring shoots or spears. The plant in its wild form is short, branched and prostrate. Domesticated specie is vigorous, strongly erect 3 or more feet in height, grows from a woody root crown. Flowers are 1-4 in axils with cladodes or branches and are bell shaped, yellowish green with round berries turning red when ripe. Seeds are 1- 9 in the berry and berries size is ¼ to 3/8 inches. Seeds take 12-14 days in warm houses and 20-30 days in the open culture for sprouting or germination. The plant is unisexual and di-oecious in behaviour with Male and female plants growing on different plants. Underground portion or the Crown consists of Rhizomes, fleshy and fibrous roots. Fleshy roots function for absorption and storage of plant food materials. Fibrous roots arise from fleshy roots. It is the stored food material during dormancy which is utilized for the growth of marketable spears harvested during spring.

Asparagus Crop Improvement : Asparagus crop improvement is very much there in USA right since 1913 when Norton adopted different methods for its improvement. The main goals then were breeding for development of Male hybrids, fusarium and rust resistance, improved spear quality, polyploidy breeding and high yielding ability and adaptability. Male Asparagus plants yield more and live longer than female plants and comparatively more disease resistant. In America development of Asparagus for rust resistance led to the development of cultivars like Washington, Martha, Martha Washington resistant.

Varieties

International market for Asparagus is in demand for both green and blanched spears. Blanched spears are white in colour and not a separate variety. Blanching is the cultural technique of bringing white spears into the market. Asparagus Rust which had threatened Asparagus culture in USA during 1900 forced for the breeding of new hybrids. New hybrids and varietal development programme of Asparagus envisaged improved spear qualities, High yield ability and adaptability to different regions, disease resistance. Apart from this, Asparagus is a crop of very high Climacteric nature, undergoes senescence after harvest. Varietal development of hybrids must conform to the post harvest handling requirements for longer shelf life and capable of resistance to post harvest diseases i.e Bacterial soft rot (caused by *Erwinia carotovora*) and Fusarium rot. Common hybrids of Asparagus produced for International trade all over the world are :-

COUNTRY	VARIETY
UNITED STATES	Martha Washington, Mary Washington, Reading Giant, Palmetto, Colossal Argenteul, Mammoth White, UC-157 (F1), UC-157(F2), Jersey Giant, Jersey Centennial, Greenwich, Backlim.
SPAIN	Cito, Desto, Aneto
POLAND	Lucullus 1813 J (White), Lymbrass 126 (White)
FRANCE	Diane, Junon, Minerve, Larac, Alexandra, Marionet, Dorbonne-4 Lorella, Valliere Rocault (White), Dornonne vert, Jacq ma verte, Verte de California (Green)
NEWZEALAND	Astora, Tramea
INDIA	SL-831, Perfection
NETHERLANDS	Franklin F1 Hybrid (Male Hybrid)

Asparagus crop culture is meant to obtain highest yield standards through cultural management and the varietal selection is the major part. Amongst all the varieties, Jersey Giant and Jersey Centennial which are genetically the best clones and parent specific are the ones in demand. Further Male clones have been found to be the highest yielders. As such vegetative propagation should aim at producing only the male clones developed through genetical marking of the DNA. All male Asparagus varieties are now revolutionizing the breeding and growing programs of fresh Asparagus, because a light harvest from Male seeds is possible next year. However roots in only Male varieties develop a year later. This has been made possible only because of Male Hybrids which are vigorous, now dominating the Asparagus seed market. Male Asparagus Hybrids produce thicker edible spears. The two important varieties bred are Jersey Knight Improved and Franklin F1 Hybrid.

WORLD PRODUCTION AND SEASONS

Cultivation of Asparagus has very much increased in the world over in the past century. Estimated area during 1966 was 1,40,000 ha world wide (Ehlert and Seelig, 1966) with 40% in Europe, 31% in North America, 14% in Asia and 15% in South America and Africa. Asparagus as fresh and canned vegetable is arriving from allover the globe to the consuming markets.

Production of Asparagus in the European countries reported by ITC Geneva Survey report 1987 shows the production in U.K, France, Netherlands as 62800-65900 tonnes and further an import of 31539 tones during 1986 detailed as under —

COUNTRY	PRODUCTION IN TONNES
United Kingdom	500 – 700
France	53000 –55000
Netherlands	9300 –10200

ASPARAGUS IMPORTS INTO EUROPEAN UNION

MAIN SEASON CROP : March - July

Importing country	1982	1983	1984	1985	1986	Import share
F.R. Germany	14482	17819	17758	20619	21966	69.6 %
Switzerland	4154	4500	4868	5098	5571	17.7 %
Belgium	609	644	1132	1373	1417	4.5 %
United Kingdom	499	612	865	897	1276	4.0 %
France	35	198	285	661	550	1.7 %
Austria	124	227	275	308	377	1.2 %
Netherlands	98	128	83	283	247	0.8 %
Denmark	27	33	42	73	73	0.2 %
Sweden	43	43	53	64	62	0.2 %
Norway	11	13	19	19	-	-
OFF SEASON CROP : September -February						
F.R. Germany	5098	5068	7612	7158	-	-
Switzerland	1759	1606	2611	2408	-	-
Netherlands	63	36	136	103	-	-
United Kingdom	277	422	518	-	-	-
France	79	148	270	-	-	-

PRODUCTION SEASONS

Depending on the country location either in the Northern or Southern hemisphere or on the Equator the seasons vary. However the production of Asparagus spears starts immediately with the onset of the spring or rise in temperatures. The main season in both the hemispheres are thus February to June and October to February. Crops of the Northern hemisphere from February to June is termed as the Main season and that from October coming from the southern hemisphere countries as the Off season. Countries like Thailand and Malaysia, Indonesia have spread their production for all the calendar months of the year. Israel has evolved its own production season for Asparagus spears from October – February, only to get the best market prices. India lying under the same geographic locational conditions as Thailand, Malaysia and Indonesia has the bigger potential for Asparagus cultivation both as main season from February to June and also as off season from June to December, take the cultivation que from Thailand and Indonesia from June till December. Another feature of Asparagus culture is the density plantations raising so as to increase yields per unit area. The marketable seasons of the exporting countries are as under—

COUNTRY	PRODUCTION SEASON
Argentina	August – December
Australia	September - December
Chile	September – December
Equador	November – February
Guatemala	November-February & July August
Israel	October – April
Indonesia	September – December
Malaysia	January – December
Newzealand	October – December
Mexico	December-March &July-September
Peru	August – March
South Africa	October – December
Thailand	January – December
Zimbabwe	August – September
India	February – June

Fresh green Spear production seasons for different countries of the world for Asparagus are different for both the hemispheres. Interesting feature which is observed is continuous crop supplies by Malaysia and Thailand from January till December. Israel has a different season for the supply of fresh green spears from October to April. Interesting facts about this are that all these countries of Northern hemisphere have different geographical locations and seasons too. This has been done to get the best of the market prices in the European Union markets when Asparagus spears are much in demand and the period of production is termed as off season during these months. India can also very well adopt these cultural techniques to increase the harvesting schedule of spears from 5 months to whole year.

MARKET SURVEY OF ASPARAGUS FOR EUROPEAN MARKETS

Main consumers of Asparagus in the world are North American and the European regions. These are the regions which fall short in their production and ultimately resort to the import of Asparagus from all over the world and throughout the year. Market survey is necessitated because of the basic realities in production, consumption, export and Import from other growing regions. Asparagus is both consumed as fresh and canned with varietal preferences for Green and white or Blanched spears. Consumption patterns are different for different countries. Market survey is for the European markets only for which the statistics are available.

Product Mix : Asparagus as such is intended to be exported as a fresh vegetable, but certain countries in the EU like Sweden prefer spears in canned form. As green and white there is preference for varietal variations.

Major Asparagus Exporting Countries: Asparagus is arriving in the EU markets from all over the world and accordingly based on the seasons of production and crop availability have been classified as Main and Off season supply countries.

Main Season : Countries of the Northern hemisphere supplying the spears from February to July are the main season supplying or exporting countries. In Europe France is the largest producer of Asparagus after Spain. Other countries in the EU which produce Asparagus are Netherlands, Greece, Italy, Belgium and Hungary, Poland in North Europe. In North America United States of America is the largest producer.

Mexico from central America is the producer and Exporter. In Asian countries Indonesia, Thailand, Malaysia are the other producers and exporters of Asparagus to the EU.

Off Season : Countries like Peru, Chile, Australia, New Zealand are the other countries supplying Off season Asparagus crop to the EU. Another notable feature in this trade is the re-export of the crop. In this category Netherlands, France, Belgium, Spain import as well re-export it to EU member countries.

Quality and Packaging : Imports of fresh fruits and vegetables to the EU are covered under the ECE standards for quality and packaging. Products from European, Mediterranian countries must meet these standards. Products from other countries must also satisfy these standards and tolerance limits in grade, quality standards.

Tariff Structure : For Asparagus import arrivals in the EU the tariff recovery is based on three different periods in the year which have been specified keeping in view production period in the EU and the interests of the EU farmers. Though entry tariff for all countries and regions into EU has been specified to be 16 % but the periods are as :-

- 1st February to 14th August
- 15th August to 30th September
- 1st October to 31st January

Price Analysis : Price analysis compared for two different periods i.e one reported by ITC (report 1987) and the second based on Market News Service report No. 37-1995. Average monthly rates per kilo of Asparagus CIF basis in the EU markets prevailed as under —

COUNTRY	RATE / KILO IN USD
United Kingdom	3.80
France	3.80
F.R. Germany	3.14
Belgium	3.34
Netherlands	2.04
Switzerland	3.27

Rates as per MNS Report 37 of 1995 in Indian Rupees

Denmark	203.00
France	225.00
Germany	172.00 to 207.00
Switzerland	173.00 to 195.00
United Kingdom	228.00

Conclusions

In view of the expanding trade of Asparagus in the EU and other world countries every small developing country holds a promise for the export of this crop to the EU and stand in competition with other established exporters. EU import figures for 1987 signify 40,000 tonnes of import from ten different countries and with a promising annual growth. Other importing countries of Middle East and South Pacific region like Japan, Singapore, Hongkong, Korea are very near to India provide very good opportunities for the development, marketing of this crop and earning very remunerative returns. Survey further points out that FR Germany is the biggest importer in the EU followed by Switzerland, Belgium, United Kingdom, France.

COMPOSITION AND NUTRITIONAL VALUE

Asparagus is a very healthy and nutritive crop and provides all the necessary ingredients with very low calorific value, which all the more makes it more valuable and highly nutritious food. Asparagus contains a good supply of Folate (Folic acid) as well as some vitamin C and in green Asparagus some beta-carotene. Asparagus also contains phyto-chemical Glutathione which has antioxidant and anti-carcinogenic properties. Analysis of 100 gms boiled spears is as under —

COMPONENT	UNIT	COMPOSITION
Water	gms	93.6
Energy	Cal	20
Protein	%	2.2
Fat	%	0.2
Carbohydrate	%	2.9
Fibre	%	0.7
Ash	%	0.4
Calcium	mgms	21
Potassium	mgms	183
Magnesium	mgms	20
Vitamin A	IU	900
Vitamin C	mgms	26
Thiamin		0.16
Riboflavin		0.18
Niacin		1.4

For body energy provision it contains very vital vitamins like Thiamin, Riboflavin and Niacin in natural form. Nutritionally this is the best vegetable for persons on low sodium diet. Consumption of Asparagus provides an alkaline reaction in the body. WholehealthMD.com has also reported for the nutritional qualities of Asparagus, that one cup of cooked asparagus contains 43 calories of energy, 0.6 gm total fat (0.1 gm saturated fat, 0.2 g polyunsaturated fat and no mono-unsaturated fat), 2.9 gm dietary fiber, 5 gm protein, 7 gm carbohydrates, 20 mg sodium, 0.2 mg Thiamin, 263 mg Folate, 19 mg Vitamin C.

ASPARAGUS CULTURE

Soil: Asparagus is a lover of well drained soils. As such Loam or Sandy loam soils are the best choice for its culture. Sandy soils produce very early crops. Soil should be free from pest and disease infestations. Constitution of the soil should not have clay component more than 20% in it. Soil pH range is 6.5-7.5 which is very optimum. Asparagus plants show a very high tolerance to salinity.

Soil Preparation: To raise a very good plantation, because the crop is to stand for 10-15 years, soil preparation as of requirements should be very thorough to eliminate the process of weedings and water stagnation. For a very pulverised structure, the soil should be worked with mould-board ploughs, harrows, tillers. Preparation must start during winters before actual plantation. Work up the soil to a depth of 25-30 cms or 1 ft. Good preparation of the soil ensures better plant foothold and deeper root system development. Roots of Asparagus go as deep as 18-20 ft, so as this ensures its own feeding and less dependence on manuring and fertilization.

Plant Propagation : During the earliest periods the propagation of Asparagus culture was through seeds. But with the passage of time and new technologies coming up and the requirement of uniform crop growth in the fields more emphasis was laid on genetically pure clones. In this the basic emphasis is laid on the yield and disease resistance in the developed hybrids. Further clones propagated vegetatively take shorter duration to production period, parental characteristics are maintained, eliminates high risk of young age. Tissue culture technique allows production of virus free clones.

Varietal and parental genetic purity is well assured from tissue cultured plants. This ensures propagation of only male clones from the male parent materials. As such reliance is suggested only for the tissue cultured clones and from a reliable source.

Plantation : The plantation starts at a time when it is the maximum growing period which is after the fall or with the onset of spring and then after the monsoon is over. One year old seedlings or rooted vegetative cuttings or tissue cultured plants are planted. Seedlings or clones are required to be disinfected with THIRAM or BENOMYL before planting. The plants need to be fixed in the soil 6-8 inches below the soil surface for good spear emergence. Plantation density per unit area must be decided before the take up of the plantation. Recommended plant spacing are 6-12 inches from plant to plant and 4 ft to 5 ft from row to row. Closer plantings has been observed to promote higher and earlier yields in the plant. There is very little evidence that spacing affects the size of the spears, if sufficient moisture and nutrients are available.

Plant Density and Spacing

DISTANCE			
Plant to Plant		Row to Row (in Feet)	
	4.00	4.50	5.00
		Plant Density	
6 inches	21780	19360	17424
9 inches	14520	12906	11616
12 inches	10890	9680	8712

DEVELOPMENT OF CROWN

In the Asparagus culture a good developed crown is the basis of high production. As such preparation for a good crown development should start with the selection of the variety and the parent whether male or female. Development of the crown starts itself with the development of the roots in the seedling or the cutting. Seedlings when 2-3 month old are hardened in poly bags 12x 21 cm size containing a mixture of loam, FYM, Peat in equal proportions. Seedlings thus lifted are carefully placed in poly-bags by spreading the root system to all directions. Then these are fixed and placed for another 6-8 month period for proper growth in the bags. At the time of plantation the poly bags are torn off and the earth ball is placed intact in the already prepared pits or holes, taking care these plants have enough roots or well developed fleshy crown with a large number of buds.

CULTURAL OPERATIONS

Manuring and Fertilization : Due to its very deep penetrating root system going up to 18-20 ft, it has been observed that Asparagus removes very less nutrients from top soil and takes care of its requirements from well developed roots only. However practices vary with region to region and country to country. In the present days much emphasis is being laid on Organic cultured crop which commands a premium price in the market. In this culture recommended manures are Poultry manure, Oil cakes, Compost and FYM. There is no application of chemical fertilizers as such. Manuring is done at the time of plantation @ 20 tones / acre FYM supplemented with 40 kg P_2O_5 and 100 kg. K_2O. There after the general belief for fertilizer application is NPK 8:15:15 is 400 kg, Ammonium Nitrate 350 kg, Ammonium sulphate 60 kg.

Weeding : Weeding is the vital operation which is taken care at the time of soil preparation to keep the soil free from weeds. After the plantation is over, the inter-space and the beds are kept weed free for the first three years. Also hoeing of the beds keeps the beds in pulverized condition for air movement in the root zone and also for drainage of water.

Earthing-up : The operation needs to be taken up with each weeding. In areas of very low or freezing winters the plants become dry, but in areas of mild winters the plants turn yellow and stunted. It is this time when the plants be headed back or cut back at the ground level so as to provide full rest period to develop the crown. This operation also prepares the plants for production of spears desired at harvest maturity. There are two well defined systems when the plant is pruned or headed back. The systems are :

1. Taiwanese system
2. Modified Taiwanese system

The systems have the difference only in complete pruning of the plant and keeping 2-3 well developed fonds or shoots intact.

Irrigation: Annual requirement of water for the Asparagus crop has been reckoned to be 35 inches, however much depends on different climatic parameters like Climate and evapo-transpiration ratio, soil retention capacity, Rainfall. Crop production is directly related to soil water availability. Under Indian conditions with the onset of spring season dry hot spells of weather start, thus making it all the more necessary for irrigating the crop frequently to avoid water stress. Since this is the time when production is at its peak. Now a days the modified irrigation systems like sprinkler and drip irrigation can be deployed for water economy as well optimum water level in the root zone.

Plant Protection : There are no apparent insect damages to the crop in India. The main insect pests of Asparagus observed are Asparagus beetle and Cutworms. Main diseases of Asparagus are Rust, Fusarium rot. Both diseases in the Pre-harvest culture are of little importance. However the varieties presently in the market are Rust resistant which are being cultivated by the farmers. In India main culture is through seeds and not vegetative propagation. The prevalent varieties are Mary Washington which itself is a Rust resistant variety. Crown rot is caused due to ill drainage, which can be controlled by the farmer by improving the drainage itself. Otherwise there are no serious diseases which require plant protection measures.

The most serious of the diseases occur only at the post harvest stage and are *Erwinia carotovora* and Fusarium rot. Bacterial soft rot makes the spears mushy soft, develop watery soaked areas on tips. Fusarium rot appears as water soaked areas changing through yellow to brown on tips, with white or pink mold. This can be controlled by pre-cooling of the crop properly. Phytophthora rot on the spears is seen in the form of Large water soaked lesions at the cut end. These shrivel later. Control of this can be achieved by pre-cooling, maintaining proper temperature regime and early marketing of the crop.

Harvesting: Harvesting of the spears is not undertaken in the first year of its plantation but the plants let to grow vigorously and much care is given for the development of the root crown which is the ultimate basis for higher yields. The frequency of harvest period is to be properly timed for the first three years, which should have 2-3, 3-5, 6-8 weeks harvesting periods in the first three years respectively. Further Asparagus harvesting is further dependant on the climate of the area and the region. Spears are harvested before the stems open up. Harvesting is also dependant on the frequency and the harvesting system.

Taiwanese System : The plant is pruned down to the ground level and shoots that develop subsequently are immediately harvested. Harvesting is carried over a period of 3 months, followed by 3 months pause after which harvesting can again be taken up in this system.

Modified Taiwanese System: In the initial pruning 3 most developed fonds are left intact and only the new shoots are harvested. Aletrnatively, after complete pruning, 3 stems are allowed to develop fully and the remaining ones are harvested. In this system harvesting can be undertaken for 4 months.

Harvest Frequency: This depends on growing conditions and ambient temperature and can vary from twice a day to once every two days. Normally harvesting can commence in the second year. Harvesting frequency in fresh plantations be carried as given earlier. Green Asparagus is harvested when spears are 8-9 inches long above the surface and before the scale like leaves on the tip sparkle. Spear shoots should not be longer than 10-11 inches. Harvesting should be done manually, with very sharp knives preferably of V type shaped blades. The spears are cut just below the surface of the soil. Quality requirements normally specify the percentage of whiteness.

White Asparagus spears stay white because these are kept covered with soil mounds about 12-15 inches high from the soil surface. It is cut 8-10 inches below the top mound surface, when the tip of the spear has just appeared on the mound.

Harvested spears be placed upright in a crate which should be covered with wet cloth in shade, taking care that tips of the spears are well below the top edge of the crate to avoid damage due to breaking. Precaution is very necessary also to save the spears from dehydration or other quality losses through sun, wind or rain.

Quality Requirements : Buyers judge the quality of the spears on the apparent freshness, straightness, diameter, color and condition of bracts on tip. Further quality confirmation of the spears is from the tenderness, deep green colour, purplish tips that are closed and compact. Consumers prefer both thick and slender asparagus stalks. Thus stalk size is not the measure of quality, but still stalks that measure ½ inch at the base are preferable in the market. High quality Asparagus is turgid, of uniform straightness, color and closed bracts. Most buyers prefer 8-10 inches long spears with a base dia. of the spear from ½-¾ inches. Poor quality spears exhibit high fibre content, hard and woody spears. Poor quality can be controlled by devising shortest harvesting frequency intervals of 1-2 days, pre-cooling, maintaining cold chain, storing the harvested produce away from Ethylene producing items.

Yield : In the European culture and climate a reasonable yield status of 3200-3600 kg/acre has been perfected. Under Indian conditions farmers can also expect the same yields, however 2000-2500 kgs/acre with a plant density of 8000-10000 plants is good enough to get best and profitable returns. In this crop yield is a composite result of improved cultural techniques, healthy and geneticaly pure plant material, disease free clones and lastly plant density. Yield comparison of 1993 productions in various countries of the world shows USA 2940 kgs, Peru 5945 kgs., Greece 3448 kgs., Taiwan 5213 kgs per hectare.

PESTS AND DISEASES

Asparagus Beetels : There are not very many pests which attack Asparagus. Extensive damage is caused by two types of beetles which feed on the crop. These beetles are crioceris aspragi and the 12 spotted Asparagus beetle *C. duodecimpun-ctata*. Spraying the crop with Rotenone @ 300 liter per acre is considered good.

Asparagus Rust : This is caused by Puccinia asparagi and is the most serious. Small reddish yellow spots on the main stem and branches enlarge into patches until the whole plant has a reddish-brown or orange colour and ultimately the top of the plants are destroyed. Rust resistant varieties is the remedy for this. Jersey Giant and UC 157 are the slow rusting cultivars.

Fusarium Wilt : It is caused by fungus Fusarium. Wilt affected spears show brown discoloration and become wilted and stunted. The causal organism is soil borne and planting should be avoided in infected soils.

GRADE CLASSIFICATIONS

Asparagus crop spears have been classified differently in different countries, where North American grade is different from the European classification. EU regulations No. 183/64 and 1677/88, classify Asparagus grading on the basis of stem color, diameter at mid point of the spear. After size separation spears are to be tied with elastic or plastic tags or placed in plastic sleeves. Batches should be of specific weights i.e 0.5 or 1.0 or 2.0 kgs. These classifications thus are :

Colour

- White
- Voilet (Having tips of a color between pink and voilet or purple)
- Green (having tips and part of the shoot green)
- Green purple

Spear shoots should be whole, fresh, sound, free from physical damage i.e un-bruised, clean and free from external moisture, foreign taste and smells. Shoots should not be hollow, split, peeled. EU sizing by length is as under :-

- Long greater than 17 cms
- Short 12-17 cms
- Tips less than 12 cms.

North American classification grades are based on classification 1965 and is termed as Californian Agriculture Code. San-Gorgion classification is different than the Californian. These are as under:

Californian			**San-Gorgion**	
Grade	Pack Wt.	No. of Spears	Grade	No. of Spears
Colossal	904 gms	14	Mammoth	17
Jumbo	-do-	15-20	Giant	18-24
Extra Select	-do-	21-28	Fancy Regal	25-34
Select	-do-	29-42	Regal	35-52
Extra Fancy	-do-	43-67	Fancy standard	53-78
Fancy	-do-	68-100	Standard	79-115

Packaging : In North America and EU packages of wood, corrugated paper, plastic are in vogue. All designs are pyramid shaped. Packaging is water resistant, including wax corrugated cartons, with a dampened Side wall ventilation is essential to maintain storage temperature and to prevent CO_2 and Ethylene build up, 1.2 cms minimum clearance above sponge placed at the base is allowed for growth. Carton dimensions, net weights vary by market, but net weight of cartons should range between 6.8-14.5 kgs. Labelling should include country of origin, No. of bundles, Unit weight of each bundle, stem color, size, quality. Unique packaging for Asparagus has been developed by PACIFIC AG-PAK, Watsonville, California with a pack capacity of 453 gms or 1 lb per pack and named CLAMSHELL packaging. Advantages of this pack are that it contains foam at the base which keeps the spears moist and capable of growth up to 1 inch. Container is transparent, allows the customer to inspect the produce for its quality, freshness and other parameters. Lastly this is the most convenient pack for the field packaging.

Post Harvest Handling

Process of Post Harvest handling begins immediately after the harvest starts. The process envisages to reduce the post harvest losses to the minimum caused due to dehydration, quality degradation, physical problems, pathological infections for increasing shelf life of the product. This has been necessitated because of very high respiration rate of the spears. To undertake handling effectively, biological, physiological factors playing vital role must be understood. Potential post harvest losses due to physical problems can be —

- **Mechanical Damage**: Broken tips, crushed, bruised spears are the common damages and can occur at any stage of handling. This can be controlled by careful handling, reducing number of handling operations.
- **Chilling and Frost Damage** : Chilling damage is caused by extended storage of 3-4 weeks at 0-3°C. This is rarely a problem for 2 weeks or less period. For longer storage optimal temperature is 2°C. Frost damage occurs at -0.8°C or below. Damage is conspicuous with change in stem texture, color. Stems become soft, watery and grey at the tip.
- **Irregular Growth**: Opening or separation of bracts reduces visual quality. Do not lay spears on their sides and at temperature higher than optimum. This causes doubling and elongation. This loss is minimized by storage at 0°C to 2°C.

- **Ageing** :Stem bending, feathering due to over growth occures, which is a sign of senescence. This indicates improper storage conditions and delayed harvesting.

PATHOLOGICAL INFECTIONS

- **Bacterial Soft Rot** : Caused by *Erwinia carotovora.* Spear tips turn grey, pale and watery. Disease initially seen as bacterial growth at the cut end, which exudes drops. Subsequently base is softened and becomes watery. The incidence can be reduced by using a clean, sharp edged knife and by containing bacterial contamination through a 200 ppm dip in chlorinated water.
- **Fusarium Rot**: This is characterised by white mold on the bracts and spear tips, results in lesions on soft tissues and watery areas turning yellow or dark green with age. The disease occurs with longer storage at 21°C.
- **Pack House Processing** : The process is a very vital component of post harvest handling and processing. It envisages prolonging the product shelf life and making it worth for transporting and haulage to distant regions. The process involves Trimming, cleaning, Pre-cooling, grading, packing and storage.

Asparagus Handling: Asparagus immediately after harvesting from the field is brought to the Pack House or Ware house for further processing. In this process removal of field heat, washing, cleaning, grading, pre-cooling and storage is done strictly as per procedural work for the process, which are as under:-

*__Trimming to Size__ : Harvested spears are brought to the pack house for further processing. Firstly the spears are trimmed as per size classification of length, diameter and placed in respective size bundles or batches.

*__Washing and Cleaning__ : Size graded produce is then subjected to washing for the removal of dust and dirt, sand and dis-infection of pathological infections. The process is done by subjecting the spears to placing in cold water tank.

- **Pre-Cooling** : In this operation Iced water is used for pre-cooling. Precaution be observed that the spears do not come in direct contact with the Ice. Water should also contain 100-200 ppm chlorine. Spears be left in 3-5°C water for 20 minutes. Apart from this organic wash compounds in the form of powder are available in the market. These compounds are purely organic being powdered from the fruits of soapnut (*Sapindus mukorossi* or *S. detergens*).

- **Grading** :On removal of spears from water these are graded based on diameter and length.
- **Packing**: Graded spears are then bundled as per weight of ½ kgs each bundle and resorting to the strict adherance of No. of spears for each length and dia. Then these small bundles are made into bigger or carton packs.
- **Cold Storage** : Cartons are then sent to the cold store for storage and transportation to the destination. Storage temperatures for Asparagus are 1°C-2°C with 90-95 relative humidity level. With adequate air circulation, uniform temperature Asparagus can be stored for 2-3 weeks. Rapid cooling is a must for quality maintenance. Product quality declines at temperature above 5°C. This quality degradation is manifested with spear hardening, color loss, water loss, development of off flavors.

Processing : Asparagus production of spears left after fresh dispatches is subjected to brine processing and canning of the spears. This process has been necessitated because of excess production of the crop and its proper disposal and capability for later usage. This process requires cooking of the spears. Cooking or boiling of spears makes these free from dis-agreeable odors, flavors or organic compounds. Quality spears of uniform thickness and maturity are selected. These are cut to Can size, washed. For blanching spears are immersed in boiling water for 2-3 minutes. Spears are then filled into Cans with 2% Brine solution. Cans are then subjected to exhausting for 5 minutes at 85°C and sterilization process at 113°C for 30 minutes or 116°C for 20 minutes.

RECIPES

Asparagus as a vegetable is served in many forms like **Salads, Soups, Hot Dishes** or in combination with various sauces. The recipes are prepared either from cuts or tips or whole green spears. Asparagus as of its nature is tender, crisp, flavourful and nutritive. Fresh Asparagus adds delight to menu servings. It also goes with Meat, Fish, Fowl, Egg, Cheese, Macaroni. Serving hot with sauce, butter adds colour and delight to the recipe.

Other recipes can be had from French menu as spears garnished with red tomatoes, Rice, cooked Egg yolk. Dressing blanched spears with melted butter adds to juiciness of the preparation. Other Asparagus recipes are —

- Fresh Asparagus Oriental style
- Asparagus with Chive Mayonnaise or with Mustard butter or with Caper, Butter Sauce or with black butter sauce or with Dilly butter.

Asparagus spcar recipes are generally prepared by "Stir Frying, Boiling, Microwaving, Roasting methods". Prior to any recipe spears are boiled in boiling water along with a clove of Garlic, slice of Onion, Lemon wedge. This provides a better flavour to the recipe. Indian cookery recipe in sweet, sweet and sour curd, cream arc the prized in Indian system.

Medicinal Properties : Diuretic, Anti-plethoric, Anti-dropsical.

10. BABY CORN OR MAIZE

Maize or Corn is a crop of the American continent and it was introduced into India from America in about the beginning of 17th century. Right since then it has adopted the status of a indigenous crop culture at par with Wheat and Paddy. Corn is symbolised as a staple food of hilly and sub-mountain tract inhabitants of India. In the past Maize cultivation got extended to the plains of India too. Next only to wheat in importance and apart from a staple food of the poor, it has further acquired the status of an Industrial and vegetable crop too. With the introduction of Waxy, High oil bearing varieties and strains for Industry another transformation was discovered for its use as a vegetable in the kitchen. This is a birth story of the Baby Corn, which has transformed the agricultural and Industrial economy of Thailand. Corn growing for vegetable use is the very recent introduction which has proved to be very successful venture in Thailand and Taiwan. This particular strain of Corn / Maize has been very much praised and is known as Baby Corn. Demands in the European and North American markets for Baby Corn is growing and the major player for this export vegetable is Thailand with export of 22000 M. tonnes during 1994 into the EU. Though cultivation of Maize for grains, Starch, oils is gaining much importance but Maize as vegetable is not lacking behind. Snack foods, Alchohol, fodder are the other demand sectors for this crop. In all types of Maize it is the sweetness which is of prime importance in the varieties which are eaten as raw, roasted, steamed, boiled, canned, frozen cobs or grains. Extra sweetness genes were discovered in the year 1964 and since then the corn has acquired an unique status not only in the Industry but with the farmers too. Every country of the world since then has bred own varieties of Corn with extra sweetness in the kernels which is manifested in "SU, SE, Sh-2" genes.

SUGARY GENE CHARACTERISTICS

SU: Gene for sugarness and tenderness. The gene is contained in normal sugary hybrids. Hybrids contain less kernel sugar than SE and Sh-2 genes.

SE: Sugar enhancer gene which raises kernel sugar content, while retaining tenderness, creamy corn texture. Varieties containing this gene does not require isolation from other corn types. This gene has the capability to raise the kernel sugar content above the standard SU hybrids.

Sh-2: Varieties having this gene are termed as super sweet hybrids. Sugar content in the kernels is above SU gene variety. Sh-2 gene has the capability to slow down the conversion of sugar to starch and holds sweetness much longer than SU or SE hybrids. Hybrid varieties with this gene are suited for 2-3 day long ship-ping and for canning and freezing. Isolation from other corn types is a must in the cultivation of these varieties.

BOTANY

Botanical Name : *Zea mays*

Other Species:

Zea mays var *ceratina* (waxy Maize)
Zea mays var *everta* (pop corn)
Zea mays var *sachhrata* (Sweet corn)
Zea mays var *identata* (Dent corn)
Zea mays var *indurata* (Flint corn)
Zea mays var *rugosa* (Sweet corn)
Zea mays var *japonica*
Zea mays var *tunicata* (Pod corn)

Corn species differ in their physical, constitutional and breeding behaviours of each specie and has been accordingly classified as :

- Dent Corn
- Flint Corn
- Flour Corn
- Waxy Corn
- Pop Corn
- Sweet Corn
- Pod Corn
- Baby Corn

Baby Corn as a Delicacy : baby corn cobs are the only edible portion which is prized for the culinary delicacies of dish's and salads. It has the maximum sweetness when harvested. To retain this sweetness refrigerate un-husked Baby corns without loosing its sweetness for one week. Process involves Blanching of the cobs by putting the husked Baby corn into the boiling water or steam for 30-45 seconds so as the enzymatic action is stopped. Then the Husked cobs are immediately put into cold water and frozen. Remove the husk and wash before cooking. Baby corn can be eaten raw as its immature size is very appealing, taste sweet, colour attractive with a crunchy and crisp texture.

- Eaten as raw for sweetness and nutrivity.
- Served as raw with Salads of Lettuce, Red cabbage, Tomato with an added flavor of celery.
- Baby corn boiled for 4-5 minutes, served as a desert with butter, salt and lime.
- Baby corn is a speciality with Soya/Tomato sauce with Noodles/Pasta/Spagheti.
- Baby corn stir fried with mixed vegetables like Capsicum, Broccoli, Carrots etc. is flavored with Onions.
- Baby corn Pakoras i.e. marination with thick paste of Black gram flour and deep fried in oil and served hot with Tomato sauce.
- Baby corn boiled and cut longitudinally and served in soups and stews.

- Baby corn pickle in Vinegar with flavors of Rosemary/ Celery/Fennel etc is a another form of its delicacy and high nutritiveness.
- Baby corn oil pickle in lemon/lime.

THAILAND SUCCESS STORY OF BABY CORN

Thailand is a country of the Asia Pacific region and is a Rice producing and consuming country. To augment the resources of its farmers through a better substitute to Paddy with higher returns, Maize was taken up as a project in real meaning for the upliftment of its economy and assuring high returns.

Credit for developing Corn as vegetable goes to the Research Scientists and the Govt. of Thailand, who were the pioneers to give lead in this type of thinking and direction. Within a span of 30 years this has become one of the continental crop as a vegetable as for as demand is concerned both in the domestic markets of Thailand and also in the export markets of Europe and the song of success can be heard in the countries of Zambia, Zimbabwe, Kenya in the African continent, who are also in the run to gain a footage in the growing demand for this crop. This is because these countries were very fast in emulating the success story of Thailand and translating the same in the perspective of their countries economy. India is also a country of the Asia Pacific region is also not behind, but the progress of emulation is too slow to convey the message of transformation, substitution of crop. India is very capable of occupying a dominant status with Baby Corn culture and marketing with Research Scientists, keen farmers, Marketers, Entrepreneurs, Industrialists and the growing consumer demand with the changing life styles of the society, can alter the scenario in record time.

ECONOMIC IMPORTANCE

Thailand basically is a Rice producing and consuming country. Rice production and culture is a loosing proposition, because of low returns in the productivity and value as well. Idea to change the cropping pattern with the substitution of another more paying crop got transformed into a very handsomely paying crop for the farmer and Industrialist as well in Thailand starting from decade of seventies. In 1974 export of Baby corn cobs was only 67 tonnes (Chamnan Chutkaew) valued at 38095 US Dollars. In the year 1992 after 18 years of culture export rose to 36761 tonnes, which was 549 times more than 1974 valued at 33 million US Dollars. Chamnan Chutkaew is of the opinion that countries of Asia Pacific region producing Maize could take a lead from the Thai experience in the promotion of Corn crop Culture, Industry, domestic consumption and boost the prospects of economy with a new hope to the farmer. Maize is adaptable to varied agro-climates, high yielding and fast growing in habit is the best crop.

PRODUCTION IN THAILAND

There are six regions of crop culture in Thailand. These are North, North East, Central, Southern, Western, Eastern. Western region produces maximum Baby corn cobs with a yield of 80277 tonnes during 1990-91. North and North-East region followed next with 25613 and 17315 tonnes of production in the same period. Central region was the lowest in its production. Baby corn has proved to be very good crop substitute for Rice cultivation. Production of Baby corn is ever increasing from 64197 tonnes during 1982-1988 to 1,29,647 tonnes during 1990-91 with per hectare productivity of 6360 kgs of cobs.

Table I. Showing Baby Corn Production and Productivity in Thailand

YEARS			
Production (in tonnes)			
1982-88	1988-89	1989-90	1990-91
64197	84309	163,501	129,647
Productivity (Kgs/ha)			
6675	6381	6908	6360

Source: Department of Agriculture Extension, Thailand. 1991.

Nutritional Value : Nutritionally Baby Corn could well be compared to Cauliflower, Cabbage, Tomato, Brinjals and Cucumbers in its nutritional values (Yodpet, 1979). However the values are presented in the Table below:-

Nutritional Value of Baby Corn (Analysis of 100 gms)

		CROPS					
Components	Unit	Baby Corn	Cauli-flower	Cabb-age	Tomato	Brinjal	Cucumber
Moisture	%	89.10	90.30	92.10	94.10	92.50	96.40
Fat	gms	0.20	0.04	0.20	0.20	0.20	0.20
Protien	gms	1.90	2.40	1.70	1.00	1.00	0.60
Carbohydrate	mgm	8.20	6.10	5.30	4.10	5.70	2.40
Ash	gms	0.06	0.80	0.70	1.60	0.60	0.40
Calcium	mgm	28.00	34.00	64.00	18.00	30.00	19.00
Phosphorous	mgm	86.00	50.00	26.00	18.00	27.00	12.00
Iron	mgm	0.10	1.00	0.70	0.80	0.60	0.10
Vitamin	iu	64.00	95.00	75.00	735.00	130.00	0.00
Thiamine	mgm	0.05	0.06	0.05	0.06	0.10	0.02
Riboflavin	mgm	0.08	0.80	0.05	0.04	0.05	0.02
Ascorbic acid	mgm	11.00	10.00	62.00	29.00	5.00	10.00
Niacin	mgm	0.03	0.70	0.30	0.60	0.60	0.10

Source : Yodpet-1979

Varieties : Baby corn varieties must conform to High yield, yellow cob colour with good row arrangement and resistance to Downy Mildew with wide adaptability characters. Prior to 1988 only open pollinated varieties were in culture. The year 1988 marked the replacement with a new hybrid of Baby corn KBTX 3501 which proved to be very promising with highest yield potential and qualifying to the demands of Industry and Export.

TECHNOLOGY TRANSFER

Full fledged mass campaign to popularise about this nutritive new vegetable crop, foreign exchange earner, Employment generator, substitute for Rice cultivation with very high returns was undertaken through Mass media of News Papers, Radio and Tele-vision, Shows, Talks, discussions. Advertising about the crop also did not lag behind. This was a great success in communicating the desired message about Baby Corn as future vegetable crop. Training in crop culture was demonstrated to the farmers. Next step in the popularisation was convincing the people for canning in the Industrial sector. Fairs, Workshops and Exhibitions further gave a wide publicity. Consumers were rallied for acceptance as vegetable through Eatery shows and demonstrations in cooking and presentation of dish variations.

Thus Baby corn culture, acceptance by the Industry and growing consumer demands both in the domestic and Export market are the live examples of 26 years concerted and dedicated efforts of all the concerned in the popularization of this new crop in the world over. Thailand Government realizing the importance was whole hog in the promotion of Baby Corn culture, marketing. Apart from this interaction, liaison with the farmers, Industry, Brokers, wholesalers, Exporters and progressive public relation campaigns for the popularization of this crop is a rare example of determined efforts and time bound success.

WHAT IS BABY CORN

Highly immature cobs or ears borne in the sheath on the regular Corn plants 4-6 days old Ears in the culms after emergence either with tassel emerged or not and harvested resulting in the collection of small ears is called "Baby Corn" with a de-husked size of the cob ranging from 2-5 inches or 6-12 cms in length. The ear is dehusked and desilked to make it a marketable product known as Baby corn. Fresh Baby corn has better flavor, crispy texture, more sweetness. Baby corn is used more like a vegetable than anything else un-like Sweet corn for grains only. The surge and demand of Baby corn over the world has grown only because of fancy, appeal for this new type of vegetable highly ranked as one of the most delicious delicacies of the Gourmet. Usage of Baby corn in the kitchen has added a special effect in preparations of various kinds and providing a gourmet touch to dish's and salads. It is a prized delicacy in the service of top class dish's. Baby corn has found much acceptance in the chinese dishes, soups, salads and dressings. The crop has gained much popularity as a vegetable of speciality with the consumers, caterers, Hotels and Restaurants at par with the marketing channels of Superstores/ markets, Importers, retailers.

Israel has devoted much attention in the breeding of Baby Corn varieties with specific emphasis on the increased number of Cobs up to 5 per plant, in order to raise the productivity per unit area and thereby the returns to the grower. HAZEERA seeds from Israel are producing the seeds of Baby corn. In North America BURPEES are the producers of the seed for this crop. Thailand has developed very good strains and is the foremost producer of Baby Corns. Other countries which has followed suit are Kenya, Zambia, Zimbabwe in the African continent and now India too. Annual growth rate in demand which is any indicative of popularity of this crop suggests concerted efforts from Indian Research and Farmers too. Breeding of two Baby corn varieties at Vivekanand Krishi Anusandhanshala, Almora, namely VL-42 and MEH-14 is a very good beginning to catch up growing market of this crop and also India's most vulnerable position in the centre. Baby corns are harvested 4-5 cms long with a base dia not exceeding 1.5 cms. Cobs of this standard are sweetest and can be eaten even raw. Baby corn is also finding its place in the processing Industry as a Brine canned and frozen product.

CROP CULTURE

Cultivation of Baby corn is very similar to Maize culture in India. The difference lies only in the harvesting of the immature cobs for the market. Other requirements remain thc same.

Soil: Maize is essentially a rich soil crop, grows best on well fertile and drained loamy that are neither too light or too heavy in its structure. Baby corn has the capability to grow in tropical or colder zones also. In India its cultivation can be extended even upto 7000 ft. m.s.l. Another commendable research which goes to the credit of Indian Research is the culture of Maize whether for grains or for Beby cobs through out the year.

Manuring and Fertilization : Corn is a very heavy feeder of nutrients as such application of NPK and micro-nutrients in the shape of chemical fertilizers in the balanced status in conformity to the soil analysis is very necessary.

Irrigation : Maize culture under Indian conditions is during the onset and culmination of monsoons signifies very high moisture status of the soil and its water requirement during growing period. Minimum of 30-35 inches of irrigation is sufficient for good growth.

Cultivation Technique : The demand for Baby corn cobs in the export market is developing very much for the organically cultured crop. For this purpose very many organisations have come up in the field for certification standards and giving the ECO-labels. This type of produce in the market and with the consumers is finding not only much acceptance being a non hazardous crop but also fetches a higher remuneration in the auction table or in the market. Organic culture recommends non application of chemical fertilizers and adherence to Bio-controls for the pest and diseases of the crop. Further emphasis is the crop culture without least damage to the environment.

Seed Rate : Seed rate is double than the Sweet corn culture i.e. 16-18 kgs /acre.

Days To Maturity : Grain or other types of Maize crops take usually 70-90 days for maturity, but in the case of Baby corn production starts from the 40th day onwards and continues till the last cob is harvested from the stalk. The process lasts for nearly 30 days from the first harvest. Total crop cultural maturity period is from 40-70 days. With the crop maturity and the supply contracts and regular market supplies maintenance, one can resort to regular crop culture for assured supplies and production.

Yield : The yield component is very much dependant on the number of cobs harvested per stalk and also the breeding behaviours of the variety. Normally a Maize plant is capable of yielding 4-5 cobs, each weighing 3 gms, will thus give an yield of 900-1000 kgs/acre.

POST HARVEST HANDLING

Maturity Index and Harvesting

Sweet corn looses its sweetness and flavor rapidly after harvesting. As such harvesting of the cobs is recommended at the desired maturity stage which is visible from the silk in case of Baby corn and also for grain cobs too. In the case of grain cobs silk when turns black brown in colour and brittle in nature. But for Baby corn the silk should be fresh and emerging. In the case of grains cobs the maturity is seen by opening the green sheath then pressing a grain with finger nail. If water juice or milk exude it is too young and if the grain contains paste it is too old. When creamy milk type liquid spurts out from the grain, it is the desired maturity index for grain cobs.

Maintenance and Preservation of Freshness: Maintenance of harvest freshness in Maize cobs and Baby corn is primarily aimed at to preserve sweetness, tenderness, texture. Process of ageing or senescence leads to conversion of sugar to starch and lowering of nutritive value. Harvest freshness maintenance is achieved by lowering or reducing the field heat of kernels and cobs. The process involved is through :-

1. *Hydro-cooling* : Corn cobs are passed or immersed in water maintained at 2°C. No sooner the core or central pith of cob attains this temperature, cobs are then packed and stored.
2. *Blast freeze cooling* : The cobs are subjected to Blast freeze air circulation over the cobs to reduce the ultimate field heat to 0-1°C.
3. *Cold storage* : Cobs harvested from the field are immediately graded and stored in cold store at a temperature of 0-2°C

For the crop quality maintenance so that Baby corn cobs does not loose sweetness, crispness, natural flavor, very high standards of crop handling requires to be observed right from the harvesting till it reaches the consumer. Crop handling immediately starts with the harvesting of the cobs. Place the harvested cobs in the plastic crates covered with moist cloth. Shift the harvested cobs to the processing unit. Harvest the cobs of the desired length and with the freshest silk on it.

De-Husking, De-Silking : This is a composite process which starts simultaneously. After removal of the husk remove the silk from the baby cobs too and place in trays.

Grading : Sizing of the baby cobs be made as per length and making lots of uniform length for creation of customer attraction. Care be observed and taken to reject the split or broken or over mature cobs.

Pre-Cooling: Immediately after grading the bulk produce be subjected to pre-cooling with blast Air so as to reduce the field heat of the cobs and for the maintenance of sweetness, which is accomplished with the reduction in temperature regime.

Packing : The pre-cooling lots can then be easily tray packed for 250 gms and providing a cling film covering on the tray to create MAP conditions. Small tradeable packs are ready for marketing. These are then put to bigger Fibre board cartons of 5 or 10 kgs each and stored in the clod store at a temperature of 3-5°C.

Potential Harvest Problems : is the maintenance of freshness, sweetness of the harvested cobs. Harvested cobs be either sent to Packhouse line immediately for processing and Pre-cooling and ultimate preparation for domestic market or Export dispatches. However for smaller quantities the cobs are immediately subjected to the process of blanching with hot and cold water and then the cobs with husk are frozen. This process saves the sweetness of the baby corn cobs.

Other problems of prime importance are freedom from mechanical damages caused at harvesting stage and handling of the Baby corn at the Packhouse, leads to secondary infections of the cobs, thereby reducing the marketable quality. Most critical problem after harvest is the problem of water loss control.

CULTIVATION ECONOMICS

Production Parameters

a. Plant density per acre	80,000 plants
b. Number of estimated cobs /stalk	4-5
c. Average weight of dressed cob	3 gms
d. Production/acre: No. of cobs x wt./cob x Density/acre	
e. Rate per kg. in Indian market	Rs 40/- per kg
f. Estimated production /acre	900-1000 kgs
g. Gross returns / acre	Rs 36,000/-
h. Cultivation and Processing costs	Rs 16,000/-
i. Net Returns	Rs 20,000/-
j. Yearly cycles/unit area	4
k. Net yearly returns	Rs 80,000/-

WORLD MARKET OF BABY CORN

Though Baby corn is a product and cuisine of Asia but Baby corn's production and markets have expanded very much to every part of the world. At present very many African countries like Zambia, Kenya etc are very much there in the production and export of Baby corn to the other parts of the world. This distinction achieved by Baby corn is because of its not being only a niche product but also a specialty item of cuisine which it could attain in the shortest possible period in the other parts of the world. With the passage of time the exports of Baby corn has grown very much, in which 76% share is that of Thailand.

PRODUCTION

Though one can produce Baby corn from standard sweet corn cultivars, but breeders have now focused their attention for producing hybrids for Baby corn only with the same characters of sweetness, crispness, flavor and colour but with an added advantage to be included with the production of 5-6 ears in one plant, which not only adds to production per unit area but also raises the economy of the farmer. The other important factors considered for Baby corn is reduction in mechanical damage. For the Indian farmer to compete in the market of Baby corn exports, a well defined planning in the present scenario when Indian farmer is already producing Baby corn but still not in a position even capable of catering the domestic needs. Then the export potential which has a definite edge in higher and very good earnings needs to be looked with a broader perspective of commercial angle for the country. In the North India, Vivekanand Institute of Research and Technology at Almora in Uttaranchal is the key center involved in the breeding of Baby corn varieties and distribution of seeds, which is yet to gather the commercial momentum. Information on Baby corn production in the world is limited because Baby corn has not been recognized as a special crop, but included in other countries of the world with Sweet corn production. However Thailand is the first country which has given a separate status for Baby corn and figures for its production, domestic sales, Exports are categorized separately as for other crops in every region of the country. Production figures of 1990 thus highlight a production of 116,600 Mts.

Exporting Countries : Major countries in the trade of export in the world are Thailand, Srilanka, Taiwan, China, Zimbabwe, Zambia, Indonesia, South Africa, Nicaragua, Costarica, Guatemala and Honduras. In the world thus Thailand is the only country which could provide an official status to Baby corn in the Export statistics, where it exported 2100 Mt's Baby corn to 22 different countries worth 1 million US$. Thai exports of the canned Baby corn during the year 1993 totalled 34,600 MT's worth 30 million US$. United states was the largest importing country during that year.

EXPORT MARKETS OF WORLD REGIONWISE

North America: The imports of fresh Baby corn are very rare in the market. Costa Rica is the largest supplier of fresh Baby corn to USA market, although it is available from local sources from California, Mexico, Guatemala and Honduras. The major import to USA is of the canned Baby corn exclusively from Thailand. During 1993 the export to USA totaled 14,310 MTs worth 12 million US$. This has been possible to Thailand because of introducing product in different weight packs. The largest pack size was 6 packs of 1.5 kg each. The price of this pack during 1994, ranged between 15-17 US$. US firms are purchasing now Baby corn in bulk packs and are canning their own way in brined form. IQF baby corn though most oftenly used in prepared foods has the smallest US market for its sale.

Europe : European countries import more of fresh baby corn than USA and the suppliers are Zimbabwe, Zambia, Kenya from Africa and Thailand, Srilanka from Asian continent. UK is the largest European market for fresh baby corn. Here in UK fresh baby corn are the retail consumers unlike USA where Restaurants are the consumers for this product. Sale figures of UK indicate that 8-10 MTs purchase is made by Supermarkets and supplies arrived year round from Thailand in packaged and loose form. The volume of imports of fresh Baby corn from Thailand to other European continent countries is $1/4^{th}$ than that of UK. Next to UK was Denmark in the import of Thai baby corn. The other importing countries of Europe are Germany, Netherlands, France.

Middle East: Market News Service reports show no import of Thai baby corn into the region. However the import data is of the re-exports to the Middle East countries from Netherlands. The other important fact was that the exports to Middle East were from the African countries via European Export agencies.

Asia : In this region the importing countries are Japan, Hongkong, Singapore, Korea, Malaysia. During the year 1993 Japan was the second biggest importer of Fresh and canned Baby corn after Malaysia, importing 136 MTs fresh and 3338 MTs canned. For a total worth of 0.21 and 4.2 million USD respectively.

Marketable Baby Corn Products:

- Fresh Baby corn
- Canned/ Brined Baby corn
- IQF (Individually quick Frozen) baby corn
- Other products i.e. baby corn in Vinegar, Baby corn Pickle in oil/Mustard flavors.

European Union Market Prices of Baby Corn in Rupees

Period : 12th October, 1995

Source : Market News Service

Importing Country	Price per kilogramme			
	Currency	Import price	Price in US$	Price in Indian Rs.
Belgium	BFR	255	8.5	297.50
Denmark	DKR	40	7.2	252.00
Finland		48.80	-	268.50
Holland	Hfl	12.94	-	284.00
Norway	NKR	85	10.3	380.50
Sweden	SKR	70	11.2	382.00
Switzerland	SFR	11.4	9.0	315.00
United Kingdom	STG	5.2	-	260.00

Exporting Countries : Thailand, Zimbabwe,

Usage : Commercial utility of corn cobs is either for grains or for vegetables. For the Industrial use the processing of corns is done for Grains, Alchohol etc. Other usages are cobs for roasting, steamed cobs, freezing of kernels in Brine and Baby corn cobs in Brine.

CORN KERNEL PROCESSING

Freezing : Most simplest, cheapest, efficient, natural way of preserving kernel is processing. The process involves no change in flavor, texture and nutritional value. However, the process envisages the principal of fast freezing and correct wrapping or packaging. The commercial process of freezing is done by machines through blast, breeze drying under Vacuum or

Cryogenic freezing. In the home scale advantage is available in the deep freezers labelled "Fast Freeze or Auto Freeze or Super cold".

Process : It involves blanching of the materials which are prepared for the process by cleaning, washing, dicing and slicing. Principal involved is retarding the action of enzymes which would otherwise cause deterioration in colour, flavor and texture.

Roasting : Sheath of the cobs is removed and the cobs are then uniformly roasted over the burning charcoal or heater or stove or kept in the pre-heated oven. For addition of taste to the cobs roasted thus are treated with lime and salt.

Steamed Cobs : Cobs alongwith green sheath are put into heated sand or boiling water and kept till the kernels get softened. When fully steamed sheath is removed, cob is brushed with lime, salt and butter and served hot.

Processed Kernels : Kernels processed either by freezing or in brine can be put to various uses i.e. soups, Salads, garnishings and dressings.

Bay corn cobs are also processed as frozen and in brine for dish preparation.

11. PEAS

Peas is a cultivated vegetable crop for its fresh grains and pods belonging to the Leguminoseae family. There are other Peas cultivars which are wholly used as vegetable for its edible pods because of no parchment membrane in the green pods and thus make it more acceptable unlike Beans as a prized vegetable of the Gourmet dish's. Peas are cultivated both for shelled Peas and for edible pods without grains and with grains. This has further two classifications and are sold in the market at much higher price in comparison to shelled Peas. Peas pods prior to the formation of grain and without the development of string are SNOW PEAS. Pea pod that breaks like a green Bean is called Snap Pea. **Snow Peas and Snap Peas** are a store house of nourishment, nutrients and energy in the form of minerals, proteins, low fat content. Peas in the dried forms are in use since ancient times. It was not until 16th century when the tender varieties were developed to be used as fresh green pods. History records that Louis XIV's courts discovered the delight of eating young fresh Peas. In the 19th century Peas were the biggest research project initiators for Mendel for laying the foundation for modern Genetics. Peas today are sold in every form i.e. fresh, frozen, canned. Fresh Snow Peas and Snap Peas are in much demand in the European and American continent apart from South East Asia and the Pacifics but for India. Snow Peas are also called Sugar Peas and Chinese pea pods because they are picked before the seeds have developed in the pod. Probably Snow Peas were developed in the 16th century in Holland. Sugar Snaps are a cross between Snow Peas and the Green Peas and were developed only in 1970's. Sugar snaps have plump edible pods filled with extremely sweet and tender peas. Peas are bush or trellising and having vine type crop features.

Shelled or Green Peas : are in demand in the market only for their grains and not the pods as such which are not consumed. Mostly the world over the demand for shelled type of Pea varieties is for fresh shelled, frozen, dehydrated and canned products in the market.

Edible Pod Peas: are an ingredient of oriental dish's, command highest market price. These are eaten raw or snapped pods and cooked like green Beans. Pods of Snow Peas are tender, fleshy, thick, sweet, with wonderful flavour. Young pods are string-less, brittle and succulent. Snow Peas in China are typically called **China Peas** because of their extensive usage in stir fried Chinese dish's. These edible pod Peas are generally known in the world as Snow Peas or Snap Peas. However these are also known with other names as **Sugar Peas / Mangetout / China Peas**. Plants of the edible pod groups are both dwarf or semi-determinate and indeterminate type with very many varieties coming in the market. These groupings of edible Pea pods are as :-

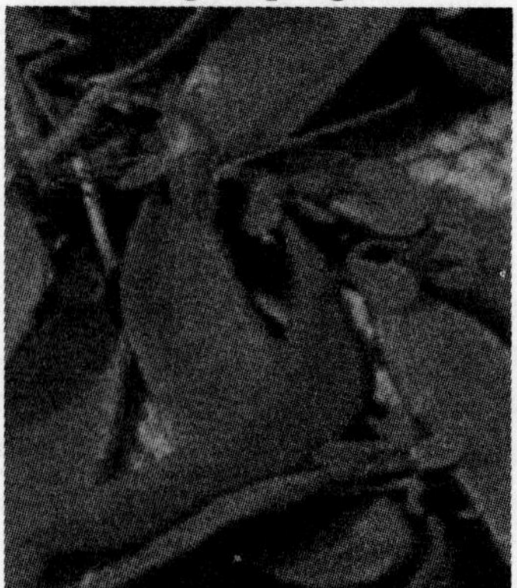

Snow Peas

Snow Peas : This group has fleshy succulent highly sweet pods without any parchment membrane, harvested before the development of the string and sold in the commercial market as immature pods without peas in it. Snow Peas is a cultivar of Pisum sativum var. macrocarpom. These can be cooked whole unlike Broad Beans with full nutritive value contained in the pod. These are ultra sweet in taste.

Snap Peas : Breeding of Snap Peas goes to the credit of US American breeder Calvin Lamborn, who brought out this group of Peas in the year 1960 in the market after a successful cross between Snow Pea and a tightly podded Pea with thick walls. Peas pods in this group are that break or snap like a green Bean. Pods are thick walled, sweet, edible without strings. The pods of this group are consumed whole with fleshy, sweet tasty Peas. This group of Pea varieties are different in different parts of the world. In the USA, Europe the varieties are very specific. Now in India the breeders have also bred the Snap Pea at the Punjab Agricultural University in the edible Pod group and is known as **Mithi Phali.** World varieties of the edible Pea pod group are —

AMERICAN VARIETIES

SNAP PEAS

1. **Bush Snapper** : Crop available after 58 days, 23-25 inches high vines, grow without sports. Early and productive pods are 3¾-4 inches long, containing 7-8 green Peas. Variety is resistant to Powdery mildew.
2. **Snappy** : 63 day fruiting crop. Productive vines, resistant to powdery mildew. Pods are 4-4¾ inch long and are much longer than Sugar Snap, contain 8-9 Peas. Vines grow 5 ½ -6 feet tall and require support.
3. **Sugar Snap** : Maturing in 70 days, vines attain a height of 6 ft. Pods are ½ -3 inches long, resistant to Pea wilt. Pods are round, slightly curved, medium green, thick, fleshy, sweet, crunchy.
4. **Sugar Bon** : 56 days crop maturity, with 18-24 inches high vines, pods are 2-3 long, resistant to powdery mildew.
5. **Snow Bird**
6. **Sugar Daddy**
7. **Climbing Sugar Snap**

8. **Cascadia**

Snap Peas

SNOW PEA

Oregon Giant or Snowman : 65 day maturity crop.

Oregon Sugar Pod I : Both Oregon Sugar pod and Oregon Sugar Pod II have been bred by Dr. James Bagget, of Oregon State University. Difference between both is of disease resistance. Snow Pea varieties produce 1 pod at each node, where as Oregon Sugar Pod bears 2 pods per node. Oregon Sugar Pod II is resistant to Pea enation virus and powdery mildew

Oregon Sugar Pod II : 68 DAYS maturity, with 28 inches tall vines. Each cluster bears 2 pods. Variety is resistant to Pea enation virus, common wilt, powdery mildew. Super variety for freezing. The variety is an improvement over Oregon Sugar Pod which bears only one pod. Pods are 3-4 inches long.

Mammoth Melting Sugar : ready for sale in 75 days from sowing. Vines are 4 ft. high with 4 – 4½ inches long pods and ¾ inches wide thick, succulent pods, uniform in productivity and wilt resistant. Variety good for freezing.

Yakumo or Sumo : matures in 90 days with a very high vine going up to 2 – 2 ½ meters in height. Pods start bearing at the 19th node.

DWARF WHITE SUGAR

SHORT 'N' SWEET

EUROPEAN VARIETIES

SNAP PEA GROUP

1. **Honey Pod** : Earliest and the sweetest, producing smaller but more pods.
2. **Sugar Ann** : Vines are 18-30 inches tall, very productive with big fully edible pods, with juicy sweet flesh.
3. **Sugar Gem** : Medium sized sweet succulent pods.
4. **Sugar Snap** : 5-6 ft. high vines. Pods are tender, tasty, with thick juicy pod walls free from parchment membrane.

SNOW PEA GROUP

1. **Norli Mangetout** : Peas are early and productive, crisp sweet pods, free from in-edible fibrous membrane / parchment.
2. **Oregon Sugar Pod II**: 3 ft. tall vines. Heavy crop-ping with vigorous, medium long – flat pods, ¾ inches wide, borne in clusters of two's.

INDIAN VARIETIES

1. **Mithi Phali** : 90 day maturity, with a growing height of 4-4 ½ ft. vines. Pods are light green, 4-5 inches long. It is highly sweet and rich in proteins and total sugars. Average yield is 47 qtls / acre, which is 2 times more than the shelled varieties.

Peas Varieties : Pea varieties are both dwarf and tall. Dwarf varieties attain a height of 15-24 inches and do not require any support. In tall varieties Peas are supported by bamboos or ropes etc. In the shelled Pea group the Indian varieties are :-

Arkal, Azad P1, Early Badger, PH 1, Meteor, Early Giant, Harbhajan, Asauji, P23, Pant Matar 2, Pant Sabzi Matar 3, Jawahar Matar 54, Jawahar Matar 3, Jawahar Matar 4, BL Matar 7, Narendar Sabzi Matar, Arka Ajit, New Line Perfection, Bonneville, Lincoln, Pant Uphar, IP 3, Jawahar Matar 1, Jawahar Matar 2, Jawahar Matar 83, Type 163, Matar 88, Madhu, Kalyanpur 123, BL Matar 3, NP 29, Kinnauri, Silvia, Multifreezer, Swaranrekha, BL Matar 1, Jayanti, Punjab-87, Punjab-88, Mithiphali (Snap Pea), Ageta –6.

GROWING PERIOD, ENVIRONMENT AND CULTURE

Growing period differs with each variety and on location factors. Environ-mental requirement cold climate at the growing period. For high yields every factor in proper management and culture of the crop is very necessary to be adopted. Temperature is the vital in its good crop and pod growth. Plant spacing is the second, where plants spaced at 4 inch apart in a line get very good space and exposure, hence branch more profusely to give higher pod yields.

Climate : Growing period varies with the variety and the group, however Pea crop does best under cool conditions. Peas are very sensitive to heat and temperature above 30°C. Ideal growth temperatures for Peas are 24°C and 7°C. Average temperature of 15-18°C are the best for Pea culture. Generally the Pea crop takes 60-100 days for its maturity in the field. Peas grow under cold atmospheric conditions.

Seed Germination: Seeds germinate when day temperatures are 50-60°F or 22°C and the minimum 5°C. Ideal pH range of the soil and the nursery bed, soil is 5.8-6.8.

Sowing and Spacing : Peas for its culture are cultivated during the winter season and the cultivation starts from the beginning of October. There are generally 3 crops taken which are early season, main crop and the late season. Spacing for the early season crop is kept at 3 inches between the seeds and is 4 inches for the main and late season crops. Distance between the rows is 1 ft. In the Western countries the row distance is increased to 1.5 to 2 ft. and to keep the density of the crop as the same two rows are adjusted side by side with a spacing of 4 inches between each row thus forming a single row. Seeds are sown in well prepared fields 2 inches deep in rows. Plant density thereby comes to nearly 1,30,000-1,70,000 per acre. Spacing between plants of 4 inches has been reported to be the best for high productive yields as this allows more space for vines to grow side shoots without congestion. Yields are 23% more than the plants spaced at 2-3 inches in a row.

Indeterminate Varieties: All the indeterminate varieties require trellising or anchorage or training for good crop yields and exposure of the vine to complete photosynthesis. Nearly all the varieties of Snow or Snap Peas attain a maximum height of 1-1½ meter but for Yakumo which goes up to 2 ½ meters. Thus these varieties does need proper trellising for support and exposure.

Seed Rate : Seed rate for the early season crop is 45 kgs. per acre where as it is 30-35 kgs per acre for the main season and late varieties.

Fertilization and Nutrition: Main nutrient required by the Peas is Phosphorous. Nitrogen is fed to the plant by the rhizobacter in the root nodules, however the demand depends on the soil fertility status. Potassium is required in very small quantities. For good supply of nitrogen by the bacteria in the root nodules, inoculation of the seeds is a must. Too much application or application beyond recommendations may result in vegetative growth of vines, poor flowering and pod set. Snow Peas are particularly susceptible to high nitrogen applications.

Insect Pests : Insect pests most likely to attack Pea crop are Onion thrips, Bud worms mites, aphids and cutworms.

Diseases : Pea enation virus and powdery mildew are the worst diseases of Peas. Pea enation virus has the symptoms of vine curling, produce small, yellowish or no pods at all. Powdery mildew shows incrustation of white powdery mass on the leaves. Other diseases which may attack Pea crop are Bacterial Blight, foot rot.

Harvesting : Harvesting is the most critical part for Peas as conditions of harvest are very different for shelled, Snow Peas and Snap Peas. Consumer requires Snow Peas of the maximum length or size and without the development of seeds. The pods must be flat and contain no cross fibers, so that the pod is eaten as whole. This necessitates harvesting of pods every 3 days from the vines. Normally it takes - 10 days for Snow Pea pod maturity for harvest from pollination to full pod size. For Snap Peas the pods are fully developed and are like a normal green or shelled Pea crop.

Yield : The yield of the green pods for shelled ranges between 45 – 75 qtls / acre. In case of edible pod variety Mithiphali it is 47 qtls / acre.

Tips for High Yields

- Inoculate the Pea seeds before sowing with Pea bacterium for formation of more root nodules, which aid in the fixation of nitrogen in the roots. Bacterium is both available commercially and also from the soils which had the Pea crop in them. Take the soil from down below at 6-9 inches and paste it over the seeds by making seeds wet and then sowing. Seeds are cultured with bacterium by dissolving 250 gm of Pea Rhizobia bacterium in 10% gur solution. Seeds are mixed with this and let to dry and then sown.

- Regular harvesting of Pea pods whether for shelling or for Snow Peas adds to more crop yields.
- Observe for the occurrence of powdery mildew and Pea venation virus affected plants and remove these and bury in the soil.

Home Freezing of Peas and Pea Pods: Home freezing is a slow process than Industrial freezing. Food quality is very negligibly lower to the IQF. Steps involved in the Home freezing process are the same as for IQF or Industrial process but for slow freezing in the Refrigerator. Sooner a product is frozen, its qualities will be better preserved. This is because all foods contain water, which expand and destroy the cell structure. Faster a food is frozen smaller will be the water crystals in the cell. To achieve this objective in the Home Fridge in comparison to the commercial IQF process, processed and cooled products are placed in very thin layers in the tray in the freezing chamber, which has a temperature of -20°C. Product quality in Home freezing can further be brought at par by cutting / dicing / slicing of the products during the product preparation. This size reduction of the product leads to further quickening the process of freezing. Various steps involved in the Home Freezing process are as under:

- Washing
- Trimming
- Cutting to size
- Blanching
- Cooling

Washing: Vegetables, Fruits collected from the garden or purchased from the market are placed in water for washing, which is also a recommended step for removal of the field heat of the product.

Trimming : Washed products are then cleaned and trim-med for the next step of cutting / slicing / dicing / cubing of the product. The trimmed vegetables and fruits are again washed.

Cutting : The product received for cutting/dicing/slicing/ cubing is done with accordingly to reduce it in size so that time taken for Blanching and freezing is cut down further for making the quality frozen product.

Blanching : This is a process deployed for the control of Enzymatic actions in the product which start the chemical reactions that provide the product a dis-agreeable odour or flavour or color. Fruits may or may not be blanched. Fruits intended for cooking may be blanched. Process provides for better flavour and texture. Blanching involves the subjecting of the product to boiling for a very short period.

Cooling: Blanched product is then immediately submerged in Ice cold water to cool it down. After the product has been cooled down in the cold water, the water is discarded and the product is subjected to drip drying or making it free from water. Immediately after this the product is spread in trays /plates in thin layers and placed in the deep freeze chamber of the refrigerator for freezing. The process of freezing is very fast in the Frost Free Fridges than in the older type of Fridge versions. After 3-4 hours of freezing in the deep freeze chamber the product the same physical texture as that for the Industrial processed ones and is now ready for packaging.

Equipment for Freezing

- Stainless steel Pan of a good capacity 5-6 liter
- Immersing container or Wire basket or Cheese cloth
- Ice cold water
- Clock or Watch
- Strainer or Colander for drainage of cold water.
- Blotting paper / Tissue paper / News papers for drip drying the cold product
- Trays for spreading the product in thin layers
- Packing materials i.e PP bags, Plastic box's, Wax cartons, Foils or Shaped foil dishes or containers.

Storage : Peas if kept at room temperature, half of the sugar gets converted into starch within a period of few hours.

Refrigeration or storing at low temperatures conserves the sweetness, freshness and tender texture and nutritional value of the pods. Storage quality of the Peas and Pea pods lies in the tender pods with fresh grains and sweetness. Overripe pods or dull coloured be not selected for purchase. Shelled Peas in the market though look good which may save time and labour but are mealy and not very sweet because of the rapidly sugar conversion under high temperatures. For proper storage place the Pea pods in a plastic perforated bag in the refrigerator. Do not keep the shelled peas but immediately cook.

Blanching and Freezing : Storage of the shelled peas is best achieved by blanching the grains and freezing and packing these at home. The process is very simple and not even time consuming. Put the shelled peas in the boiling water in a colander and add a pinch of salt to bring up the green colour of the peas. No sooner the water starts boiling again, note down the time and keep it for 2-3 minutes in the boiling water. Then immediately sieve the contents or peas, letting the water to be thrown. Thus put the boiled peas immediately in ice cold water and keep it for some time till the peas have cooled down to near icing temperature. Put the peas on a paper or cloth for rinsing excess moisture and then put in a tray and keep in the freezing chamber for 3-4 hours by which time the peas are in frozen state. Take out these and pack in plastic bags and keep in the freezing chamber till to be used.

Nutritional Value : One cup of cooked peas provides the nutrition of 134 calories of energy, 0.4 gm total fat (0.1 gm saturated fat, no mono-unsaturated fat, 0.2 gm poly-unsaturated fat), 8.8 gm dietary fiber, 9 gm protein, 25 gm carbohydrate, 5 mg sodium, 0.4 mg thiamin, 3.2 mg Niacin, 0.4 mg vitamin B6, 23 mg vitamin C, 101 mcg folate, 62 mg magnesium, 0.8 mg manganese, 187 mg phosphorous.

Packaging: Perfect frozen product quality is all dependant on the type of packaging them irrespective of the type of Freezer/ Fridge. Inadequate or incorrect packaging or wrapping may lead to product exposure to very low temperature of Air in the freezer, which will dry it out and making it tasteless, un-pleasant, cross flavored. Freezer pack material are moisture and vapour proof, resistant to breakage or cracking at low temperature. Range of Pack materials for Home freezing are :-

- **Polyethylene** Plastic bags of 100 - 120 gauge.

- **Plastic Boxes** preferably rectangular in shape with top hard lid.
- **Waxed Cartons** : These are rigid waxed cartons capable of withstanding water and moisture entry into the carton and product.
- **Foils** : These are Aluminum foils of heavy duty, shaped in dishes, small trays and box's.

Take out the plates / Trays from the Fridge with the product fully frozen and invert it on a News paper / Blotting or Tissue paper. Hammer it gently with the hand and the product will be separated into individual pieces. Fill this frozen product in Polythene or self sealing poly bags, exhaust the air from the bag and now staple it or seal it and place it in the deep freezing chamber of the fridge. The product thus frozen is equally at par in its freezing qualities to that of the Industrial grades of IQF quality available in the market. The frozen products can be utilized during the lean months or at times when the products are available at very high prices in the market. House wives can make use of this technology for the smallest of the fridge 135 liter capacity with 3-5 kg of the products and up to 10-12 kg in 330 liter capacity fridges. Thus the products which find much acceptability in the homes are **Peas, Cauliflower, Carrot, Tomato Puree** etc. can be accessed without any trouble.

Sugar Pea : Edible pod Pea variety maturing in 68 days from the date of sowing. Vines are 2 ½ ft. tall and require staking for very high yields. Pods appear in clusters of two pods. Pods are 4 ½ long and ¾ inch wide, very superb for pod and shelled grains harvests. Variety is much in demand for with 5 Star Hotels, Consumer chains being a taste of the Western recipes with very sweet pod without any parchment membrane. Yield of the pods is about 40 –45 qtls / acre, but when these are harvested for shelled pods the yield is 70-80 qtls/acre. For edible pods harvest the pods without any grain in it, which carries the highest market rate at the auction table. Seeds are round and light green in colour.

Sugar Pod : This variety is also known as Mangetout or Snow Pea. This is the finest variety in its class of edible Pea pods. Vines grow 4-4 ½ ft tall in 68 days. Pods are thick succulent, crisp, sweet and delicious, 4-4½ inches long with a uniform productive habit. Pods appear in clusters of two's on the vines with 7-9 grains per pod. The variety is highly suited for edible pods and the yield is 45-50 qtls per acre and for shelled pods it is 80-85 qtls per acre. Variety grows well in cool climates. This variety is wilt resistant. Seed is white and round in its characters.

Table : Showing features of popular Pea varieties

Cultivar	Maturity (days)	Node where flowering begins	Vine height cm	Pod length cm	Peas per pod	Pods per node	Disease tolerance
Group							
Snap Peas							
Sugar Bon (dwarf)	59	11	50	7.6	7-8	1-2	1, 3
Sugar Ann	58	10	52	7.6	7-8	1-2	1
Honey Pod	70	16	51	7.6	7	2	1
Climbing Sugar Snap	75	16	125	8	7	1-2	1
Cascadia	65	-	100	7.6	-	2	1
Snow Peas							
Oregon Sugar Pod II (dwarf)	65	15	90	10	2	2	1,3,4,5
Oregon Giant	65	16	100	10-13	2.5-2.8	2	3,4,6
Yakumo	90	19	200-250	13	2	1-2	Nil
Sumo	68	13	100-150	13	3	1-2	Nil
Mammoth Melting Sugar	69	16	97	10	1	1	Nil

Disease tolerance :

1. Common wilt 2. Near wilt 3. Powdrey mildew 4. Enation mosaic virus 5. Wetern pea streak virus 6. Fusarium

Source : NSW Agriculture Flemington Markets Reporting Service, Luke Jewell Seeds Pty Ltd and Sunland Seeds Pvt. Ltd Australia

12. SARSON AS POT HERB

Sarson is a member of Cruciferae family and the genus Brassica and is popular the world over for its medicinal and culinary, vegetable qualities from the seeds, greens, oil. The residual content of the seeds is used as cattle feed and as organic matter in the crop culture. Sarson is known under different names in the world i.e Mustard, Rape, Black / Brown / White or yellow Mustard, Rape seed, Sarepta Mustard, Palai Rape, Ethiopian Rape, Abyssinian Mustard, Rai etc. All these names confirm a definite variation in main plant characters for seed size, colour, leaf, stem, plant size, and further more on chemical composition of oil etc. have attained a specie specific distinction in the Brassica genus.

In India various species of Sarson (Brassica) are known under different names in various regions of the country as Lahi, Laha, Rape, Rai, Toria, Mustard. Sarson is a main oilseed crop of Indian continent and is available as Brown / Black / Yellow Mustard in the market. The Rai oil has a little content of volatile oil, odor of which is very pungent. After oil extraction from the seeds the residual content is the oil cake used as a source of animal feed. This cake with a leftover content of oil is further subjected to oil extraction by solvent extraction method and the cake then is oil free or de-oiled cake. This de-oiled cake is again used as a cattle feed or as an organic manure for the crops. Sarson leaves and stems as greens are being used and consumed as a pot herb. Sarson leaves are having varied other uses in the form of seasoning, flavoring of recipes. Crushed or whole seeds of Rai and Mustard are used in Pickles for enhancement of flavours. Ayurveda has given other uses of Sarson as a home or village medicine.

Culture of all Brassicas is undertaken during winters in India and is a tradition and a best source of Bee flora for Honey and Bee Pollen collection. In Punjab this has proved to be boon to the farmers after the decade of 1960's when it placed the state of Punjab as a big producer of Honey. Mustard Honey is lightish yellow in colour and very tasty.

Origin : of Sarson is unknown but some believe it as from the Mediterranean centre with secondary centre in the North East.

Botany : Brassica is a very broad based genus includes in it apart from Sarson other members of **COLE** group of crops such as Cauliflower, Cabbage, Knol-khol, Broccoli, Brussels sprouts, Radish, Turnip.

Botanical Name	Local/Common/English Name
Brassica nigra	Black / Brown Mustard
Brassica juncea	Brown Mustard, Rai, Indian Mustard
Brassica rugosus	Palai Rape
Brassica alba syn. *B. hirta* syn. *Sinapsis alba*	White or yellow Mustard
Brassica besseriana	Sarepta Mustard / Asl-Rai, Brown Mustard
Brassica carinata	Ethiopian Rape seed/Abyssinian Mustard
Brassica compestris	
Brassica rapa	
Brassica japonica	
Brassica napus	

Plant : is a perennial herb but grown as annual or bi-ennial, erect with long branch's with tap root system. Flowers appear in racemes. Plant height varies for each specie and ranges between 2-6 ft. Plant is fairly resistant to droughts

Climate : Hardy cool season plant, grows best in conditions of monthly average temperature of 15-18°C. Thrives best in areas with hot days and cool nights. Rainfall of 3.5-17 cms is quite sufficient for good crop. Best growth temperature for all sarson species is 5.5 – 27°C.

Soil : Culture of Brassicas suits very much for light, sandy loam, deep rich fertile soils. Grows in soils with a very wide range of pH of 4.5 – 8.2.

Seed Rate : is generally the same in all the species which ranges between 1.6-2.0 kgs/acre. In Britain the seed rate for heavy and light soils are 4.8 and 5.8 kgs/acre. Seeds are sown maintaining a plant to plant distance between 4-6 inches with a row to row distance of 9-18 inches.

Sowing Time : starts with the onset of winters in India and spring in Europe. Certain varieties have been bred with very short period of 90 days maturity, which have a special place in the Crop rotation for having 2 crops in the same season thereby also increasing the crop intensity. Culture of Toria before the main crop of Wheat is one such example. Sarson is also sown as a mixed crop in the rainfed areas to have the benefit of weather vagaries and avoid crop failure.

Fertilization : For a good crop of Sarson seeds for oil, the recommended fertilizer doses are :-

- N 20 – 30 kgs/acre
- P 40 – 60 kgs/acre
- K 20 – 30 kgs/acre

Yield : Sarson is mainly cultivated for seeds and oil production. However it is also cultivated around the Metropolitan and Urban cities for its greens. In Metros and Urban areas the demand for greens during the winters as a vegetable is very high. Yield levels for seeds and greens are as under :-

Yield Kgs/Ha

	Species	Country USA	Srilanka	India
SEEDS	*B. nigra*	1100	400	—
	B. juncea	1100	—	900-1235
	B. alba	900	—	—
GREENS	*B. juncea*	—	—	12000
	B. alba	—	—	37000

Harvesting : *Brassica nigra* flowers after 45 days and crop for harvest is ready in another 6-7 weeks time. *Brassica juncea* has 40-60 days growing time depending on variety and weather conditions. *Brassica alba* (White or Yellow Mustard) requires 3-4 months for its maturity.

For seeds the crop is harvested when pods fully mature. Plants are kept in the fields after harvest for 3-5 days and then thrashing and winnowing processes are undertaken to bring out seeds.

Oil Extraction : is extracted by any of the 3 processes of extraction whether it is "Expeller, Rotary Mill, Hydraulic Press". 60-65% residue left after oil extraction is the oil cake. Oil cake has also a left over of oil percentage in it, which is further extracted by solvent extraction process and making residue as oil free and is termed as De-oiled cake.

Post Harvest Management : is needed only in the case of Green harvest for vegetable, which requires to maintain the freshness of leaves till it is consumed. Present system involves packing in small bundles in the field itself and then putting water over it or Hydro-cooling with very small quantity of water and then carrying it to the auction/marketing centres in the Urban/ Metro markets in gunny sheet packs. By the time it reaches consumer there is good amount of quality deterioration due to moisture loss. This also leads to nutrient loss and ageing of the fresh vegetable. Post harvest management of greens for quality delivery, moisture, nutrient loss savings, processing at the village level and value addition to the greens for a remunerative price accrual in the market is very necessary. The process thus is —

- **Washing:** Harvested greens are washed fully in free clear water in big vats.
- **Pre-Cooling :** Washed produce is left in water for a good amount of time so that the field heat is reduced to the maximum and there is very little moisture loss in the transporting of the produce. In the Western countries Pre-cooling for the greens is done, by giving Ice cover over the loader/crates or packages which ensures freshness, quality delivery to the consumer with an added advantage that the bundles in packs can further be kept for 1-2 days without any loss of quality in it.
- **Packaging :** Precooled produce is then packed in consumer packs or bundles in a defined weight of 250 or 500 gms or 1 kg. To add further value to it, this can be further put in polypacks to save from the moisture loss and increasing the shelf life till it reaches consumer.

USES

Seeds

- Whole or crushed seeds as condiment in pickles, paste.
- Oil extraction
- Seed residues as cake for cattle feed and manure.
- Seed Tea for sore throat, Gargles (GREIVE – 1931)

Oil

- For frying, illumination, lubricant, soap manufacture.
- White Mustard oil for Mayonnaise dressing.

Leaves and Stems

- Pot herb, seasoning, Salads, raw eating.

- Leaves, stems are canned and frozen.
- Greens are used for soups and stews.

Medicinal

- Mustards are Di-uretic, Emetic, Rubefacient, Stimulant.
- Mustard plaster for Arthritis, Rheumatism
- Hot water on bruised seeds acts as a stimulant for Foot bath, cold, Headache.
- Oil stimulates hair growth.
- Chinese eat leaves in soups for inflammation, Bladder and Haemmorrhage. Seed or oil is used both internally or externally for Cancers, growth of abdomen, spleen, stomach, throat, Uterus.
- Koreans use seeds for abscesses, Cold, Lumbago, Rheumatism, stomach disorders.

13. WATER CRESS

Botanical Name : *Nasturtium officinale*

Other Names: Water cress, Brooklime, Brown cress, Cresson, Cress, True water cress, Nasturtium.

Habitat : Water cress is a perennial found growing in running water streams or near cool shallow springs or spring holes. In India this is a perennial herb and is used as a pot herb in the villages. In North India it is found everywhere in the cool regions along side of the running streams and water courses. Water cress is found as wild growth in the perennial streams of Himachal Pradesh. Jammu and Kashmir, Uttaranchal states of North India.

Plant : Leaf is compound with leaflets oval and dark green in colour. Stems are fleshy and succulent with white fibrous roots present on the nodes. The plant presents a very attractive dark green foliage with mustard odor when rubbed. It flowers from April till June with white flowers very attractive for Bee's.

Cultivation : The plant can be cultivated both by seeds and vegetatively from root and stem cuttings. Seeds are sown during the month of March-April in already well prepared seed beds of very light soil. Seeds are sown superficially on the seed bed and irrigation water is applied in abundance to create a very high moisture regime. Germination of seeds takes place in 3-4 days.

From then onwards the status of irrigation is kept regular for a good crop. It requires very generous application of irrigation water and seed bed always moist. Roots or stem cuttings are planted in pot culture and nurtured as nursery plants, till are capable of being transplanted in the seed bed with a daily regular supply of water to the growing plants. Rooted plants are then planted at 6 inches from plant to plant and 1 ft row to row distance. The plants spread in a very short period with abundance growth of leaves and multiplication of stems in a highly abundant moist seed beds.

Constitution : The plant is very rich source of Vitamins, Minerals and Proteins having Vitamins like A, B-1, B-2, C, D,E, K,P. Other constituents contained in Water cress are Arginine, Aspartic acid, Beta-carotene, Biotin, Folacin, Glutamic acid, Glycine, Histidine, Iso-leucine, Lysine, Methionine, Pantothenic acid, Phenylalanine, Serine, Threonine, Tryptophan, Tyrosine, Valine.

Properties : The herb is edible as a poor man's pot herb, food and is valued as medicinal in nature. The herb is Antiscor-butic, Depurative, Di-uretic, Expectorant, Purgative, Hypoglycaemic, Odontalgic, Stimulant, Tonic and Stomachic. Culpepper states that Water cress juice or bruised leaves free the face from blotches, spots, blemish's when applied as a lotion.

Medicinal Value : Constitution and medicinal values are self indicative of the usefulness of the herb. Freshly pressed juice of the herb is used internally and externally in the treatment of chest and kidney complaints. Medicinal poultice of the herb leaves is said to be an effective treatment for glandular tumours or Lymphatic swellings and chronic irritations and inflammations of the skin.

Usage : Water cress is mostly used in India as a pot herb. However elsewhere in the Western countries its usage is in the form of Salads, Sand witches, dressings and in cottage cheese.

14. BEANS

Beans belong to the Class- Dicotyledonous and are the members of Family Leguminoseae. This group of vegetables is the most important as plant yields Bean pods, Seeds (small and large), fixes nitrogen of the air in the soil and enriches the soil status. In the large grain category of Beans are the Snap, Faba (Broad Beans), Lima, Kidney Beans. In the small seed types of Beans are the

Black gram, Green gram, Red gram, Horse gram, Cowpea, Pigeon Pea, Cluster Bean, Chickling Vetch, Purple Vetch and Kulth Beans, Soya Bean. All these Beans are very heavy yielders of oils too and Soya Bean is one of the commercial oil crops in the genus Glycine. Beans belong to sub-family Papilionaceae. More prominent Genus and species which are under cultivation in India for pods, seeds, fodder, green manure and oils are as under. There are very many other species which are specific to their areas of origin and value only as a wild plant for seeds or fodder etc. Edible Beans popular for crop culture are French, Haricot, Runner Beans.

Edible Beans : Green snap Beans, Yellow wax beans, Scarlet runner beans are the ones in demand as edible Bean crops. Green Beans are a rich source of nutrients carbohydrates, proteins, vitamin C, folate, iron, and a little beta-carotene if deep green in colour. Fresh beans can be classified in two groups :-

- **Edible pod Beans** – Green, Snap and Chinese beans
- **Shelled beans** - Lima and Fava beans

Edible Pod Beans : In this edible Bean pod group more popular edible Bean pods are "Chinese Long Beans, Haricot, Italian green Beans, Purple wax Beans, Scarlet runner Beans, Snap Beans / String Beans/ Kidney Beans (Lobia).

Shelled Beans : In this group the popular species are Cranberry Beans, Faba Beans (Broad Beans), Lima Beans and Soya Beans.

BOTANY

Family : Leguminoseae

Phaseolus vulgaris	Kidney Bean, Farash Bean, Vilaiti Sem
Phaseolus lunatus syn. *P. limensis*	Lima Bean
P. aconitifolius syn. *Vigna aconitifolius*	Moth Bean
P. aureus syn. *Vigna radiatus*	Moong, Green gram
P. calcaratus syn. *Vigna umbellate*	Rice Bean
P. coccineus syn. *P. multiflorus*	Multiflora/Scarlet/Runner bean
P. mungo syn. *P. radiatus*	Black gram, Urad
P. lathyroides syn. *P. erectus*	Phasemy Bean
P. lunatus syn. *P. limensis*	Sieva Bean / Lima Bean

P. trilobatus syn. *P. trilobus*	Mugani or Mugam Bean
Vicia faba	Broad bean, Windsor Bean
Vicia benghalensis	Purple Vetch
Vigna sinensis syn. *V. catajang*	Cowpea, Lobia
Vigna unguiculata syn. *V. sinensis, catajang*	Cowpea, Lobia
Dolichos lablab syn. *Lablab vulgaris*	Kulth Bean, Horse gram
Cyamosis tetragonoloba	Cluster Bean, Guar
Lensc ulinaris syn. *L. esculenta*	Lentil, Masar, Masoor
Cajanus cajan syn. *C. indicus*	Pigeon Pea, Red gram, Tur, Arhar
Conavalia ensiformis	Horse / Jack / Bara Bean

BEANS VARIETIES

- **Lobia :** Pusa Phalguni, Pusa Dofasali, Pusa Barsati, Pusa Rituraj, Pusa Komal, Phillipine Early, P – 85-2, P – 426, Narendar Lobia 1, Type 1,
- **Lima Beans / Sem :** Pusa Early Prolific, Pusa Sem 2, Pusa Sem 3, Rajani, Kalyanpur Type 1, Kalyanpur Type 2, HD 1, HD 22, HD 18, JDL 53, JDL 85, HA 3
- **French Beans :** TL Dwarf 1, Pant Anupma, Pusa Parvati, Contender, Kentucky Wonder, UPF 191, Pant Bean 2, Black Queen, SBM, BL Bean 12, Arka Komal, Selection 2, Premier, Bush Beans Gold Wax, Pole Bean Blue Lake
- **Guar :** Pusa Sadabahar, Pusa Makhmali, Pusa Navbahar

CULTURAL REQUIREMENTS OF BEANS

Soil : Bean crops love rich light loamy well drained soils least reactive to compaction. Sandy soils are thus much better for Bean culture, having neutral or slightly acidic pH levels.

Time of Sowing : Beans are a hot weather crop. Cultivation of Beans start in the temperate regions from the month of March till June. In the hilly areas of North West Indian states of Jammu and Kashmir, Himachal Pradesh, Uttaranchal Beans are cultivated during hot weather or during summers as in the Temperate regions of the world. In the plains of India too these are cultivated during summers.

Seed Rate : Seed rates are different for determinate and Indeterminate types. Determinate types are dwarf in character. For dwarf varieties seed rate is 30 kg per acre and it is 12 – 15 kg for indeterminate types.

Seed Treatment : seeds are treated with rhizobial cultures for better germination and good get up of the crop. Seeds are also treated with fungicides and insecticides.

Days to Maturity : Maturity of the crop is a dependant factor on the type of crop is aimed for harvesting. Beans cultivated for greens have different days of maturity for Snap, Faba and Lima Beans. The periods are 60-75 days for Snap, 130-150 for Faba or Broad beans, 80-100 days for Lima Beans. But when the crop is for dry seeds another 30 days be added to the green vegetable crop. For green manure application the period is all the more shorter.

Nutrient Requirement : Manuring and fertilization is the most important aspect in the Bean culture and all the other vegetable crops. Crop responses for good growth are un-predictable because of soil, climate, seeding/plant density etc. Adequate levels of nutrient supply must be guaranteed with the soil tests, for obtaining top class high crop yields whether green pods or dry seeds. Manurial schedules should be based on the crop density, taking higher the density higher the schedule applications. Recommended application rates per acre for NPK are 40 : 25-60 : 25 kg. Sulphur and Zinc application at many occasions have given very good yield results. Micro-nutrients like Magnesium, Boron may also find application for better yields of greens.

Irrigations : Irrigation is restricted up to 2 weeks from the date of sowing. Too much water at seeding may slow down growth and increase root rots in Beans. During the rest of the period keep moisture levels at near availability of moisture to the growing Beans. Major moisture stress periods during culture of Beans are the blooming and Pod formation. It is best advisable to irrigate crop only during the morning hours when soil temperature is not too high.

Yield : Yield status is also different for both the types. Yield for green pods ranges between 30-40 qtl. per acre for dwarf or determinate types and 40-60 qtl. per acre for indeterminate types. Yield for dry seeds is 4-5 and 6-8 qtl. per acre for determinate and indeterminate types respectively.

Post Harvest Management: For green pod production post harvest management starts, immediately from harvesting. Green pods are cooled down for fresh market sales and stored at 4-7°C with RH at 95%. Fresh green pods can thus be stored for 7–10 days. These temperatures may cause chilling injury but are good for small period storage. Beans prefer hydro cooling, because it cools down the produce immediately and protects green pods from wilting and shriveling. Though Beans can also be cooled by vacuum and forced air cooling systems but these are not preferred. Processing for post harvest management and storage is different for every type of Bean crop.

Packaging : After washing, cooling processing the Green Bean pods are graded and packaging is done. Under Indian conditions the crop has not gained any popularity for post harvest management and the Bean pods are brought to the market as such without cooling and processing in ordinary gunny packs, where these get damaged more and farmer gets poorest returns for his hard labour. It is thus recommended that Beans of all types are processed and then packed in plastic crates of 10-15 kg capacity or else pods can be put in consumer packs of ½ and 1 kg packs contained in 10-15 kg cartons.

Pests and Diseases : Anthracnose, Angular Leaf spot, Fusarium blight, Mosaic virus are the diseases which are of concern for Beans. Use Bavistin treatment of seeds and spray of crop with 0.1% bavistin spray and treatment. Disease resistant varieties are very much popular now a days.

Mites, Blister Beetle, Bean bugs are the common pests in Bean crop. Spray of Malathion, Endosulphan, Metacystox is recommended.

BROAD BEANS

Local Name :	Bakhala
English Name :	Broad Bean or Windsor Beans
Botanical Name :	*Vicia faba* var *major*
	Vicia faba var *minor*
Family :	Leguminoseae
Sub-family :	Papilionaceae

Broad Beans are an ancient, cultivated, *Vicia faba* var *minor* small seeded relative of the garden Broad Beans *Vicia faba* var *major*. Broad Beans origin is regarded to be in West or Central Asia. In whole of Europe Broad Beans is used as a livestock feed. In the Middle East it is a special winter crop used and consumed as a vegetable. Faba Bean looks like a smaller version of Broad bean and grows upright from 1-1.5 mtr high. This is an annual legume with hollow stems, strong tap root system, compound leaves, with large white flowers having purple or black markings. Flowers are produced in clusters bearing 3-4 pods. Pods are 1-2 cms in width and 6-15 cms long.

WHY BROAD BEAN CULTURE

Broad Bean is cultivated for various usages like cover crop, Green Manure, Silage, Forage, Hay and Vegetable. In the clayey soils it has a special importance as soil loosening through its deep rooted system. Green Manuring is done at the time when Broad Beans starts blossoming. Broad Beans are a very good supplement of high protein requirement both in the humans and animals too. This contains 24-30% protein. Humans consume it as a vegetable in fresh/dried or canned forms. It is very good livestock feed for poultry, pigs, calves and lactating dairy animals. Broad Bean has thus been classified as the best contract crop for markets as a vegetable and Livestock feed.

Where Produced in World Countries

UK is the largest producer of Broad Beans as a winter and spring crop. In Canada commercial culture of Broad Beans started in 1972. Other countries of the world where this is being cultivated are China, India, Middle East countries, North African and South American countries.

CULTURE

Soil : Broad beans grow on a wide range of soils from loam to clayey and under a variety of drainage conditions. Soil pH for culture is between 4.5 – 8.3. Low pH leads to delayed development of nodules, thus preventing atmospheric Nitrogen fixation. Broad beans does not tolerate extended periods of saturated soils and drought.

Sowing : Fields are well prepared to make well pulverized beds for sowing directly or drilling the seeds with a drill. Mostly this is cultivated as a winter crop, however in UK this is cultivated both as a winter and spring crop. Hot dry spells during the cultivation cause crop wilting. This can also be cultivated in assured irrigation conditions. Faba broad beans withstand heavy frost. Broad beans is a very shy crop should not follow Oil seed crops or legumes in the rotation. Seeds are sown 1-3 inch deep in the soil.

Bacterial Inoculation: Broad beans being a legume crop be treated with the Faba bean inoculum only. Rhizobial inoculums for Peas, Alfalfa, clovers are not effective for Broad beans.

Interculture : Broad beans do not compete with weeds or grasses in the culture, as such young plants be given a proper attention for hoeing and weeding in the earlier stages of plant growth.

Days to Maturity: Maturity period is determined by the type of the output whether for green pods as vegetable or green manure or seed or silage etc. However the general period of is as under :—

- Days to blossoming 45-60 days
- Days to pod formation 80-115 days

Seed Rate : Seed rate differs with the type of culture it is intended for. For vegetables the plant spacing is done 6x6 inches and 12 inches from line to line. Seed rate is very high or double that of for vegetable culture. For forage or fodder or hay and silage usage the seed rate is generally 50-60 kgs /acre.

Pests and Diseases: Broad beans attract predatory wasps and Lady beetles, because of the flowing nectar in the flowers.

Pests : Weevils, Aphids and Root Knot Nematodes are the most known pests of the Broad beans

Diseases : Asochyta Blight causes stem, leaf and Pod blotching or spotting.

Sclerotinia stem blight : which is common in Sunflower, Rapeseed and Peas causes damage to Broad beans. **Fusarium Root Rot** : attacks legumes and oil seed crops and Broad beans is no exception to this attack. **Bean Yellow Mosaic Virus** : has been reported in Canada to cause damage to Broad beans.

Yield : As a green manure crop the dry biomass Broad beans supply to the soil is 2.1-4.6 tonnes /acre with a nitrogen fixation of 37- 90 kgs/acre.

15. CUCURBIT CROPS

Family Cucurbitaceae : Contains a very good number of vegetable crops such as Melons, Cantaloupes, Gourds, Squashes, Cucumbers, Gherkins / Cornichons and Pumpkins. The family is also well known as Gourd family. These are annual crops and grown as vegetables mainly but for Melons and Cantaloupes which are used as ripe fruits as desserts. These all are hot season crops and grow as vines or trailers. In this group only two species *Mimusops dioica* or *Mamordica dioica* and *Cucumis melo* var *agrestis* are perennials. All these are Native to different parts of the world like Asia, Africa, Eastern USA, Central America and Mexico.

HISTORY OF WORLD'S LARGEST FRUIT

Archeological evidences indicate that New World Indians were cultivating Squash's and Gourds 8000 years before. These Indians cultivated these fruits for their seeds rather than the flesh of the Gourds. These Gourd members are low in calories, high in fiber, rich in vitamin A. These fruits have very many uses as Salads, fried / boiled / steamed / pickled / candied / dried or baked or made into breads. Cucurbitaceae family is known as the family of Gourds, has nearly 100 genera and 700 species of fruits. It contains in it Pumpkins, Gourds, Melons, Squash, Cucumbers. In addition these fruits have been used since time immemorial for the creation of beauty artifacts and Musical instruments and also as water jugs by Indian saints on their long sojourns. Luffas of the family make very good sponges.

World's largest fruit has weighed in this family. This could be possible only with the World Confederation of Pumpkins, which with its formation started the competition and invited the entries from participant farmers. Family Cucurbitaceae in it contains an array of different shapes, sizes, colours of fruits and of one of the biggest sizes. In the Pumpkin Federation show of 1993, it exhibited a Pumpkin weighing 377 kg and a Squash weighing 317 kg. In the year 1994 at Elgin Ontario, competition of Gourd Olympics, it recorded this time a giant sized Squash weighing 408 kg above the Pumpkin. Confederation in one of its yearly shows at Clarence New York in the Year 1996, made another remarkable achievement in the weight of Pumpkin with 481 kg. These competitions has created an awareness of record breaks of weight for Pumpkins, when it got new record weighings of 516 kg and 605 kg Pumpkins in the year 2000 and 2002 held at Canfield Ohio and Manchester New Hampshire respectively.

Cucurbita pepo : (*C. pepo* ssp. *Ovifera* var *ozarkana*) a native of Ozark region of Arkansas is supposed to be the ancestor of Summer squash and gourds.

Sechium Edule : Gourd from Mexico cultivated since pre-columbian days is Chayote. Chayote fruits are single seeded unlike their other family members with many seeds are hence planted. Chayote fruits when mature are bitter in taste as such these are used raw or immature.

Luffa or Sponge Gourds : Of this sponge group native to Asia are the species L. aegyptica having a smooth outer rind and *L. acutangula*, having prominent longitudinal ridges.

Melons : Include some of the juiciest, sweetest, delicious fruits of the family and are *Cucumis melo*. The other members of *Cucumis* are the Cucumbers. The other gourds of this group are *C. disaceous, C. metuliferus*. Water melons arc derived from wild ancestor *Citrulus lanatus* var. *citroides*

Gourds : This group of cucurbit vegetables include Ash gourd, Bitter gourds, Sweet gourd, Snake gourd, Sponge gourd, Round gourd, Small gourd.

Bitter Gourds : is native to the old world and is also known as bitter melon (*Mamordica charantia*). Apart from its culinary value the fruits are used in folk medicine for curing of ailments like chronic colitis, leprosy and dysentery gouts, coughs, severe menstrual disorders. Bitter taste is due to tetarcyclic terpenes called Cucurbitacins. Though toxic in higher doses or concentrations but possess beneficial qualities in the cure of tumours.

Snake Gourd : Snake gourd (*Trichosanthes cucumerina*) and Wax gourd (*Benincasa hispida*) are Asian gourds. The name itself implies that the fruits are very long serpent like reaching upto 6 ft. in length.

Courgettes or Marrows : Courgettes or French Courgettes have become increasingly popular in recent years. Fruits do not grow very large, but are harvested at a very young stage and cooked un-peeled. Harvest maturity is judged when thumb nail very easily pierce the skin. Smaller marrows are preferred very much and Zucchini is the most favourite in the continental cucurbit vegetables. Young fruits of Zucchini are used both as a

Salad and vegetable. Older fruits are allowed to ripen for jams. Zucchini is a Bush plant, need little attention. Plants bear both Male and Female flowers. Where female flowers fall off, is a sign that natural pollination is not there. Insects, Bees do the pollination in these. Artificial pollination is also resorted in case of early and assured fruiting by hands. Male flowers are picked, petals are taken away and stamens are rubbed over the female flower before noon. This ensures pollination of the flowers.

Zucchini Brimmer

Zucchini Burpee Hybrid

Melons : This group includes in itself Water Melon, Musk Melon or Cantaloupes. Except for Water melons all Melons resemble winter squashes in Structure they have a thick flesh with a central seed filled cavity. Principal difference between Squashes and Melons is that Melons are used as desserts and squashes as vegetables. Melons are sweet and juicy. Only distinction between Musk Melon and Cantaloupes is that of the smooth and netted rind or skin of the fruit. The fruits of this group are very sweet, juicy, cool and refreshing. These fruits love low humidity levels and a very good sunshine during its growth. This sunshine favours development of sugar content, flavour and solidity in the fruits.

Cantaloupes are rounded fruits of small **(Kharbooza)** and bigger size **(Sarda)** or **Afghani Kharbooza**. Sarda is oblong in shape and is a fruit of the sub-temperate or temperate culture conditions. In India Hot dry conditions for the culture of Sarda are available in the Kinnaur District of Himachal Pradesh. This crop takes nearly 100-110 days for maturity in the field. Sarda fruits at present are being imported into India from Afghanistan. Figures for the import not available, however this is the one fruit which can be cultivated in Himachal can revolutionize the economy of the Kinnaur farmers, because of its huge demand in the Metro markets of Delhi, Mumbai and Chennai. Further the culture of Sarda has potential for export to the South East Asian countries of Thailand, Malaysia, Singapore and Hongkong. Cantaloupes are round, comparatively small variety of Musk melon and is usually with a netted rind / skin and highly delicate flavour and sweetness. Orange coloured Melons have exceptional quantities of beta-carotene. Central America, New Zealand and Chile are the best exporters of Melons to USA and augment the peak demands.

Water melon

Cantaloupes

Squash : This group of vegetables have Summer squash and Ash gourd. Squash is also known as Winter squash. However there is further categorization in this group to be called as **Courgettes** (Zuchhini) and Marrow. Delicious, flavourful, to be worth eaten as raw with attractive pith have been termed as **Marrow**. Summer squash is thus a Marrow in the new English nomenclature.

Squash

Cucumbers : Cucumbers were cultivated in Asia from the ancient times. Columbus introduced Cucumbers into America and are being cultivated right since then both by the colonists and the locals. Cucumbers are also called "cukes" and grow in a variety of shapes and sizes from 1 inch long Gherkins to 20 inch long cucumbers. Cucumbers are basically of two types i.e. slicing and pickling.

Classification : Various genera and species known all over the world for its culture are under very many common and local names are presented here under :-

Botanical Name	Common Name	Local Name
• *Cucumis melo*	Musk melon	Kharbooza
• *Cucumis melo* var *cantalupensis*	Cantaloupe	Kharbooza, Afghani Sarda
• *C. sativus*	Cucumber	Kheera
• *C. melo* var *utilissimus*	Cucumber	Kakri / Tar, Long Melon
• *C. angora*	Gherkin	Kheera
• *C. melo* var *agrestis*	Small Gourd	
• *C. melo* var *mamordica*		Phut, Kachra
• *C. ficifolia*	Malabar gourd	Fig leaf gourd
• *C. metuliferus*	Horned cucumber	
• *C. foertidissima*	Coyote Melon	

• *Cucurbita maxima*	Winter squash	Ash gourd, Petha
• *C. pepo*	Summer squash	Chappan Kadoo
• *C. siceraria*	Bottle gourd	Lauki
• *C. moschata*	Pumpkin	Kadoo, Kumra, Sitaphal
• *Citrulus vulgaris*	Water melon	Tarbooz
• *C. vulgaris* var *fistulosis*	Indian squash	Tinda, Round gourd
• *Trichosanthes anguina*	Snake gourd	Chachinda
• *T. dioica*		Padwal
• *Luffa aegyptica*	Sponge gourd	Torai, Ghia Tori
• *L. cylindrica*	Sponge gourd	-do-
• *L. acutangula*	Sponge Gourd	-do-
• *Mamordica charantia*	Bitter gourd	Karela, Balsam Pear
• *M. cochinchinensis*	Sweet gourd	Bhaat karela
• *M. mixta*	Sweet gourd	Bhaat karela
• *M. balsamin*	Sweet gourd	Kakora, Balsam Apple
• *Mimusops dioica* syn. *Mamordica dioica*		Jangli Karela
• *Sechium edule*	Chayote	Launku
• *Marah macrocarpus*	Wild cucumber	Kakora

Crop Culture : All the cucurbit crops are grown both as Early and Main season. In the early.crop culture the seeds are germinated first and cultured under cover along the river beds of Yamuna, Ganges, Satluj. In Himachal Pradesh, Swan river of Una District is highly fit for covered culture. Under Indian conditions covered culture is not followed unlike Cucumbers.

Western style under the poly house but in the open with a grass hedge facing the East. Germinated seeds of Melons, Cucumbers, Tinda are put in well prepared holes and watered regularly till seeds germinate and establish as fully secure growing plants. When there is rise in temperature after January the covers of grass are removed and spread over the sand for the vines to spread over it. For the open culture sowing time is different for the plains and hills. In the plains with the change of the season and rise in temperature during February- March, seeds are sown for the main crop. In the hills the season for sowing of cucurbits starts from the March month and continues till June end.

CUCURBIT CROP VARIETIES

- **Bottle Gourd :** Mahamaya, Mahalaxmi, Hybrid – 204, Hybrid – 1720, Hybrid – 320, Pusa Summer Prolific Long, Pusa Summer Prolific Round, Pusa Naveen, Pusa Sandesh, Hissar Selection Long, Hissar Selection Round, Punjab Long, Punjab Round, Punjab Komal, Selection – 1, NDBG – 1, Arka Bahar, KBG – 1, Pusa Manjri,
- **Tinda :** Arka Tinda, Punjab Tinda, Hissar Selection, Bikaneri Green, Dilpasand, S-48, A-22,
- **Pumpkin :** Arka Chandan, Arka Suryamukhi, Pusa Vishwas, Pusa Vikas, IHR 83-1-1, CO –1, Solan Badami.
- **Ashgourd :** CO -1, CO – 2, S-1, Local
- **Spongegourd :** Pusa Chikani, Pusa Nasdar, Kalyanpur Chikani, Pant Torai –1, Satputiya, Punjab Sadabahar, Rajendar Nainua, Rajendra Ashish,, Coimbatore-1, Udaipur Selection,
- **Summer Squash :** Chappan Kaddu – 1, Pusa Alankar, Punjab Chappan Kadoo-1, Australian Green, Pusa
- **Padwal :** FP-1, FP-3, FP-4, FP-5, Patar, Basti Selection-1, 2, 3, 4, HP-1, HP-3, HP-4, HP-5, Swarnarikha, Swaran Alokik
- **Kakoda :** Local
- **Kundroo :** Local
- **Snakegourd :** Local
- **Kakri :** Japanese Long Green, Straitat, Pusa Sanyog, Poinsett, Punakhira, Balamkhira, Sheetal, Kheera 75, Kheera-90, KH -3
- **Muskmelon :** Sona, Swarna, Pusa Sharbati, Pusa Rasraj, Haramadhu, Arkarajhans, Arkajeet, NDM-1, NDM-2, Punjab Sunehri, Durgapura Madhu, Lucknow Safed, Hissar Saras (MH-5), Hissar Madhur (MH-13), Punjab Rasila, Punjab Hybrid, Hara Madhu, Punjab Sunehri
- **International Muskmelon varieties :** Casaba, Crenshaw, Honeydew, Juan Canary, Persian, Santaclaus, Sharlyn.

- **Watermelon :** Swarna, Sona, Honey, Mishti, Sugar Baby, INDAM – 365, Madhu, Patanegra, Newhampshire Miggette, Ashai Yamato, Arka Manik, Durgapura Kesar, Durgapura Meetha, Improved Shipper, Special 1, Shipper
- **Cucumber :** Special Long Green, INDAM – 9531, Lucky Strike, Poinset, Balam Kheera, Kheera-90, Poinset, Punjab No.1, Kheera-75
- **Bitter Gourd :** Hybrid – 49, IA-49, IA-411, IA – 711, IA – 1124, IA –4625, Pusa Year Round, Pus Special, Pant Karela-1, Coimbatore Long, Arkaharit, Kalyanpur Bara-masi, Hissar Selection, Faizabadi Baramasi, C-96, NDBT – 9, C-96, Punjab – 14
- **Zucchini**: E 1 Hybrid (Bush), Productive (Bush), Cocozelle (Bush), Prolific (Trailer), Brimmer, Burpee, Gold Rush F1, Tiger Cross F1, Spremo F1

Sowing Time: Sowing time varies with the location whether the plains or the hills and also with the type of culture whether it has to be under covers or protected and as an Early crop culture when the sowing is resorted to purely under covers or protection from the vagaries of the season and the cold climate. Protected crop culture is done to get the Early crop for better market price.

- Early Crop under covers : November to Mid January
- Main crop of Plains : February to March
- Hills March to June

Method of Sowing : All the crops of the Cucurbit group are cultured through seed sowing. Two to four seeds are put in one whole and after full germination only one or 2 germinated seedlings are left in the hole. For early crop culture the seeds are first germinated and then seeds with protruding radicle are put in the hole and watered. Seeds for this technique are first soaked in hot water 50-55°C and left in it for 24-48 hours, which softens the seed coat and activates the germination of the seed inside. After soaking the seeds are put in moistened cotton, which is further wrapped in paper and cloth to make it a ball. This ball containing the seeds is very perfectly kept at a constant temperature of 25-28°C by keeping it in sun for whole day. During the night period the ball is again put in cotton so as to maintain the gained heat during the day. Seeds thus take only 3-4 days to germinate with this Indian technique of germination without incubators.

Cucumbers

Gherkins

Spacing and Plant Density : Vines or trailers differ in their habit of growth and spread. As such spacing between the rows and the plants also differ amongst all of these crops. Crop culture is done in the open and the spacing between the rows is maintained by the width of the beds. Seeds are planted on the edges of the bed or holes are dug out on the ridges for the pre-germinated seeds. The spacings between the rows and plant to plant are as under :-

Crop	Bed width in Feet	Distance Plant to Plant	Plant Density Per Acre
Zucchini	10	3	4800 (Bush Type)
	10	5	2400 (Trailer type)
Musk Melon	10	2	2400-2600
Water Melon	8	2	2700
	10	2	2400
Squash (Chappan)	5	1.5	5800
Bottle gourd	6	1.5	4800
	7	2	3600
	8	1.5	3600
	8	2	2700
Sponge gourd	10	2.5	1700
	10	3	1500

Ash gourd	10	2.5	1700
	10	3	1500
Kheera	4-6	2/2.5	2900-5400
Bitter gourd	4	1.5	5000-7000
Pumpkin	10	2/2.5/3	1450-2200
Round gourd/Tinda	10	2.5/3	1500-1700
Sarda Cantaloupe	10	1.5/2	2200

CULTURAL RECOMMENDATIONS

Crop	Seed Rate in Kg/ acre	Fertilizer			Maturity in days	Av. Yield in Qtls.	Seed Production in Qtls.
		N	P	K			
Musk Melon	1.5	50	25	25	70-100	50-65	1.5-2
Water Melon	1.5-2	25	16	16	95-120	72-130	1-1.5
Zucchini	1.5-2	40	15	15	50-55	65-80	-
Squash (Chappan)	2	40	4	-	60-80	60	3-4
Bottle gourd	2	30	-	-	60-70	200	3-5
Bitter gourd	2	40	20	20	55-60	35-40	2-3
Sponge gourd	2	40	20	20	70-80	35-40	1.5-2
Ash gourd	2	40	20	20	120-150	100-130	1.5-2
Tinda	2	30	15	15	60-70	25	1.5-2
Cucumber	1	25	15	15	60-65	60-70	1-1.5
Pumpkin	1.25	35	25	20	70-80	90-100	3-4
Sarda Cantaloupe	2	25	16	16	100-120	100-120	3-4

Manuring : FYM application is the vital part of the land preparation and good yield. Well rotten FYM @ 10-20 tons per acre is the minimum recommended application for all the crops. It should be applied at the time of soil preparation.

Irrigation : Crops in this group are that of the kharif season when the temperature is high enough in the atmosphere and soils get dried very fast, thereby making it highly necessary for irrigating the fields after every 4-5 days in the months of May and June. During the rainy season, irrigation be made as per the requirement of the soil and that of the crop. Crops which are grown in the river beds seldom require irrigation, because of the water table immediately below at about 3-4 feet from the root zone.

Seed Production : Cucurbit crops are cross pollinated. Seed production thus has to be undertaken with great care and precaution so as not to allow cross pollination from other varieties of the crop. For this a definite separation distance of not less than 400 mtrs is recommended, since the pollination is through Bee's and insects. The other factor is that of the unwanted or rogue plants appearing in the crop must be removed, so as the purity of the crop is maintained at its best.

Harvesting and Storage: At the time of crop harvesting crop maturity index is the best phase to determine, when it becomes worthy of harvest for best colour, appearance and acceptability in the market for higher prices. Maturity index varies with each crop. Vegetable type cucurbit crops are harvested at a very tender stage, when the seeds are still fleshy and fruits small. In the Melons the only clue to ripeness is the condition of the rind. Second sign of ripeness is the fruity fragrance Melons are plucked at mature green stage for far off markets and half slit stage for local markets. Kheera, Kakri / Tar, Bottle gourd, Sponge gourd, Zuchhini, Bitter gourd, Summer squash (Chappan), Tinda are harvested when the fruits are very tender and seeds un-developed and fleshy. Ash gourd and Kumra/Pumpkin are harvested at full maturity. Gherkins/Cornichons are harvested for Salads and Pickling and are highly prized in the market for the smallest and tender size of the crop by the Food processors. Thus harvesting for Gherkins is done immediately after the 3rd or 4th day of appearance in the flowers when the size of the fruit ranges between 3-5 cms in length and 1- 1.5 cms in dia.

Zuchhini : Skin of this fruit is very thin and fragile, delicate enough to be punctured even with a nail. Harvested Zuchhini fruits as such should be very sound and free from any nicks, pits and bruises or soft spots. While harvesting the colour should be bright, uniform. Stem end or petiole be glossy green. Store the fruits in perforated plastic bags and at temperatures above 10° C.

Cucumbers : Slicing and pickling cucumbers(Gherkins) need to be washed thoroughly to remove dirt, dust and residual impact of pesticdes and fungicides on the fruit skin. This may be done with organic fruit wash available in the market from the soapnut powder, is very safe and healthy. Washed fruits be then rinsed and then packed in plastic bags to maintain RH at 90-95%. These bags are then kept in the fridge at temperatures above 8ºC so as to avoid any chilling injury to the fruit. The fruits thus remain in very fresh condition for more than a week.

Nutritional Value : Cucurbit fruits whether for vegetables or for desserts all are a rich source of energy supply, proteins, carbohydrates, low in fat content with the absence of cholesterol fat in them. Thus all cucurbits contain proteins, carbo-hydrates, fats, minerals like sodium, magnesium, phosphorous, manganese and dietary fibers. Nutritional values presented here under have been given by http://www.wholehealthmd.com.

Zuchhini : Nutritional value of the Zuchhini for one cup cooked is 25 cal, dietary fiber 2.2 gm, protein 2 gm, carbohydrate 5 gm, 0.3 gm fat.

Pumpkin : Apart from raw and ripened pumpkin which has all the food nutrients, seeds of the pumpkin are even more nutritious. One ounce of the hulled seeds provide 153 cal of energy and containing 13 gm fat (saturated 2.5 gm, monounsaturated 4.1 gm, polyunsaturated 6 gm), 1.1 gm dietary fiber, 7 gm protein, 5 gm carbohydrate, minerals such as sodium 5 mg, copper 0.4 mg, iron 4.3 mg, magnesium 153 mg, manganese 0.9 mg and phosphorous 333 mg.

Cucumbers : Nutritionally cucumbers are good source of dietary fiber, carbohydrates and protein and without any cholesterol content in it. One medium peeled cucumber has a nutritional value of 24 cal energy, 0.3 gm fat, 1.4 gm dietary fiber, 1 gm protein and 5 gm carbohydrate, 4 mg Sodium.

Melons : One cup of diced Water Melon has the nutritional value of 49 calories energy, 0.7 gm fat (0.1 gm saturated fat, 0.2 monounsaturated and 0.2 gm polyunsaturated fats), 0.8 gm dietary fiber, 1 gm protein and 11 gm carbohydrates, 3 mg sodium, 7.4 mg Lycopene.

Pests and Diseases : Pests and diseases of cucurbits is a serious problem in the culture of these crops. Main pests and diseases which attack the cucurbits at one or the other stage starts right from the seedlings, grown up plants, leaves and fruits. Major pests are the Red Pumpkin beetle, Aphids and Jassids, Spider Mites and Fruit flies. In the diseases major losses are due to the Powdery and Downey Mildew attacks on the leaves. Collar rot is at the seedling development stage when stem rots and shows signs of watery nature, thus not withstanding the plant weight and dies. Pythium works similarly like collar rot. For this Karathane and Mancozeb fungicides are the best to control. However seeds do need treatment with Thiram before sowing. Varieties of the crops resistant to Pest and Disease attacks are the best and the need of the day for Organic culture.

16. CHAYOTE

Chayote : This vegetable of the family Cucurbitaceae is native of South America and was introduced into North Indian Hill region. This is locally called as **Launku.** Chayote in the local language is known as Launku and is loved for its fruits and not for the roots unlike South American and African countries. Chayote is a crop of the areas with high rainfall and high Relative Humidity levels of 80-85% in the atmosphere during its growth period. It loves a day light of nearly 12 hours and growth temperature of 13-21°C Temperature above 28°C gives very high vegetative growth and incurs flower drop from the vines. Life cycle of each vine thus planted is 1 – 3 years.

Chayote is widely cultivated throughout America. This was cultivated by Aztecs long before the Spanish arrived. Later it was introduced in South America, Carribean islands, Europe. It was from here that it traveled for its culture to Asia, Africa and Australia. Chayote has very much become a fruit of international importance, because of its demand. Costa Rica is the leading trader and exporter of this crop mainly to USA and to other countries like Colombia, Guatemala, Netherlands, France, Germany, Great Britain. Imports of Chayote into USA during the years 1992-95 were 17500, 13639, 14793 and 11289 metric tons, which had a share of 89% export from Costa Rica.

CROP CULTURE

Climate: During its growth or culture period the atmosphere should be humid and hot with high precipitation. It is highly susceptible to frost damages and droughts.

Soil : Launku loves deep soils with good drainage and high organic matter content of the soil.

Chayote Fruit on Vines

Harvested Chayote

Propagation : Plants are propagated either by seeds or fruits or from the basal shoots. Seeds from the mature fruits are taken out and resorted to seeding unlike other crops. Fruits throw open sprouts during storage when the period is beyond 20-30 days, temperature in between 13-21°C with adequate moisture. Else the fruits can be made to sprout by providing similar conditions for commercial plantings. The process is done during the rainy season. Basal shoots from the growing plants are taken and treated with root hormones and planted in seed beds and provided regular mist sprays to keep good humid condition. It takes 20 – 30 days for rooting.

Planting/Sowing: Land is prepared for the plantation only for commercial culture. Otherwise Chayote is cultured in the Kitchen or Backyard gardens with one or two plants. Holes are dug out and filled with soil and organic mixture. Sprouted or germinated plants are then fixed in these holes and watered regularly till it attains good growth. For twining sticks or hedge supports are provided. In the acidic soils Lime is applied. Planting is done during rainy season.

Cultural Practices: Cultural practices for a productive culture is the weeding and hoeing, fertilization, irrigation and pest management. Chayote is a heavy feeder of plant nutrients as such fertility be maintained. Apart from FYM fertilizers 15 :15:15 or 12 :12:12 are applied at the rate of 100 kg per care. Irrigation is done when needed to maintain proper moisture and humidity level. Nematodes are the worst pests which cause rotting of the Chayote roots.

Diseases : Roots, Leaves and Fruits are the most affected from fungal diseases. Leaf spots and lesions developing on the leaves do not cause that much damage which is otherwise a negative return loss from the fungal attack on the fruits. Phoma, Mycovellosiella and Colletotrichum are the fungi which attack the Chayote / Launku fruits and make these un-worthy of marketing and also reduce market prices very much. Control as such of these fungi is paramount to the fruit protection from such damages.

Harvesting and Storage : Fruits for the export market are harvested, stored at temperatures 13°-14°C at a relative humidity of 80-90%. These remain perfectly in very good condition under these controlled temperature conditions. The fruits are well packed in poly bags and paper or fiber board cartons for far off market transport.

Harvesting : Harvesting of Chayote is generally done manually. Fruits are harvested twice weekly. Harvested fruits are collected in bins for further processing and packaging.

Yield : Each vine is capable of delivering 300 fruits during its growth period (Engels and Jaffrey, 1993). Generally 9-11 tons of yield has been recorded.

Uses and Properties : Chayote is basically used for vegetable consumption. Apart from the fruit stems, leaves and tuberous part of the adventitious roots are eaten in the South, Central American countries of Mexico, Brazil. Chayote fruits are a rich source of important amino acids like aspartic acid, glutamic acid, alanine, arginine, cistein, phenylalanine, glycine, histidine, isoleucine, methionine. Medicinal use of Chayote is also well documented for its properties, to relieve urine retention and burning during urination or to dissolve kidney stones and also as a complementary treatment for arteriosclerosis and Hypertension (Lira, 1988; Flores, 1989; Yang and Walters, 1992). Dehydrated fruits have been conserved for longer periods to be used as vegetable. Jams and sweets are other products which have been manufactured from Chayote.

17. SPINACH

Spinach is a green leaf vegetable produced throughout India, is a winter vegetable crop. Spinach is available all the year round in Indian markets and is much valued as a pot herb with every home, being a rich dietary complement of the poor.

Plant : Spinach plant produces a rosette of fleshy leaves, oval / rounded / triangular. Leaves are borne on a short stem, Stem elongates with crop maturity, producing a seed stalk with narrow pointed leaves. Spinach is a salt tolerant crop.

Classification : Spinach can be classified both on the type of leaves and the seed morphology. Based on the leaf morphology, Spinach has been classified as "Savoy" - wrinkled leaves for fresh market, "Semi- Savoy" – grown either for fresh or processing markets, "Smooth" – flat leaved is for processing and as pot herb. On the basis of the seeds, seeds are smooth and prickly.

Bolting : is a common feature in Spinach culture. Pre-mature seed stalk formation is caused by late season, high temperature, long days leading to bolting of plants. To control or keep bolting under check or minimum physiological stress, such as increased day length, high temperature, close plant spacings, inadequate fertility, irrigation must be minimized.

BOTANY

Family :	Chenopodiaceae
Botanical Name :	*Spinacea oleracea*
English Name :	Spinach
Local Name :	Palak

Spinach Varieties: WS. All Green, Pusa Jyoti, Pusa Harit, Pusa Bharati, HS-23, Jobner Green, Virginia Savoy, Pant Composite 1.

Soil : It loves soils of sandy loam nature and other soils having a well drained feature.

Climate Requirement : It is a quick maturing cool season hardy vegetable. Growth is rapid at 15-18°C, can withstand sub-zero temperatures without any plant injury. Freezing temperature only harms seedlings, plants attaining maturity stage.

Seed Germination : Seeds sown during winters, under optimum temperature conditions germinate at 7-24°C. However seed germination also takes at temperatures 2-30°C.

Method of Sowing : Spinach is sown and cultivated from seeds. Seeds are sown directly into the seed beds in well prepared fields in rows. Beds are prepared for either 2 or 4 row planting with each bed having a width of 75-150 or 95 -100 cm for 4 row planting.

Seed Rate : Seed rate per acre directly depends on the configuration of rows and may vary between 5-12 kg per acre. Higher seed rates are used for fresh market productions. 35 – 38 seeds weigh 1 gm, there by creating a density of nearly 1,75000 – 4,20,000 plants per acre.

Maturity Period : Spinach takes 30 – 55 days for fresh market and 70 – 120 days for processing crop.

Yield : Green spinach yields 8 – 17 MT per acre.

Post Harvest Handling : Spinach has a very high respiration rate and a large surface to weight ratio. Immediate cooling after harvest, to prevent rapid wilting and moisture loss is undertaken. Cooling effect or bringing down of temperature of the crop after harvest much depends on the type of cooling being undertaken. Vacuum cooling brings down 19°C temperature -3°C of the harvested crop within 10 minutes. Forced Air cooling from 18°C- 1°C is brought down in 4 hours time. Spinach stored at 0°C with 95-100 RH level has a shelf-life for the Fresh market for 10-14 days. Spinach is very sensitive to Ethylene and moderately sensitive to freezing injury after harvest.

CHAPTER - 17

IMPROVED VEGETABLE VARIETIES OF INDIA

TOMATO : Nova, Hira, Rambo Hybrid, Moti, Uday, Rupali, Ramya, Rohini, Rakshita, Naveen, Naveen – 2000, Naveen – 2000 Plus, Rajni, Ruchi, Rajat, Utpan, Vaishali, Vikas, INDAM-9501, INDAM-9502, INDAM-9801, INDAM-9802, INDAM-2101, INDAM - 2102, INDAM – 2103, INDAM – 2104, RASIM.

CHILLI : Divya, Shola, Japani, Kiran, Surya, Festival No. 5, 6, 9, 10, 21, 42, Pusa Jawala, Pusa Sadabahar, NP 46 A, HC 44 (Hissar Shakti), HC 28 (Hissar Vijay), Pant C 1, Kalyanpur Type 1, Kalyanpur Type 2, Kalyanpur Selection, Andhra Jyoti, K 5452.

CAPSICUM : California Wonder, Bharat, Mahabharat, INDAM – 3, INDAM – 5, Yellow Wonder, Arka Mohini, Arka Gaurav, Arka Basant.

BEET : Open Pollinated – DDR, Ruby Queen, Ruby Red.

CARROT : Surkhi, INDAM – 13, Early Nantes, Nantes Scarlet, Pusa Kesar, Pusa Yamdagni, Chantanny, Hissar Gorik, No.29, Selection 233, Selection 21, Imperator.

TURNIP : Pusa Sweti, Pusa Kanchan, Pusa Chandrima, Pusa Swarnima, Gloden Ball, Purple Top White Globe, Snow Ball, L-9, Punjab Safed, White 4, Red 4, Early Milan.

RADISH : Hill Queen, Mino Early Long White, Japanese White, WS – Chetki, Crimson Giant, Early Scarlet Globe, INDAM-4,7,8, Rapid Red White Tipped, White Icicle, Pusa Chetaki, Pusa Himani, Pusa Desi, Pusa Rashim, Pusa White, IHR 1-1, Japanese White, Punjab White, Kalyanpur No. 1, Jaunpuri, Arka Nishant, Chinese Pink.

BULBS

- **GARLIC:** G-1, HG-1, Yamuna Safed G1, G-288, Type 55-4
- **ONION RABI :** Pusa Ratnar, Nasik Red, Pusa White Flat, Pusa White Round, Pusa Madhavi, Early Granno, VL-3, VL-67, Agrifound Dark Red, Agrifound Light Red, Agrifound Rose, Kalyanpur Red Round, Hissar 2, Patna Red, Punjab Chaman, Punjab Selection, Punjab Red Round, Punjab 48, Udaipur 102, Udaipur 103, Vaswant 780, N 257-9-1, Arka Niketan, Arka Bindu, Arka Kalyan, N 2-4-1.
- **ONION KHARIF :** NAFED-53, Agrifound Dark Red, Pusa Red, N 2-4-1.
- **LEEK :** Mammoth Colossal, Prizetaker, London Flag, Americam Flag.

RHIZOMES

- **GINGER :** Him Selection, Rio-de-Janiero, China Type, Vynad Local, Nadia, Baruasagar, Moran, Vynad Mannonthode, Suprabha, Suruchi, Surbhi.

LETTUCE : Simpson, Black seeded, Great Lake, Ruby, Romain, Butter Head, Crisp Head.

ENDIVE : Green Curled Ruffec.

METHI / FENNUGREEK : Pusa Early Bunching, Kasoori, Metyhi No.7, Lam Selection –1.

AMARANTHUS / CHOLAI : Badi Cholai, Chhoti Cholai, Lal Cholai.

SWISS CHARD : Fordhook.

MUSTARD / SARSON : Japanese Sarson, Haryana Saag.

CHENOPODIUM / BATHUA : Pusa Bathua No. 1.

CORIANDER : Narnaul Selection, Pant Haritma, NPD 92, Pusa Selection, UD 1, 20, 21, 49, No. 5365, Punjab Katui, Faizabadi.

KALAONJI : NS 44, NS 48, Mandsaur Selection 62-1.

FENNEL : GF 1, BC 14-3-3, Finnochio.

PARSLEY : Champion, Moss Curled.

BHINDI / OKRA : Rim-Zim, Barkha, Ws – 40, Vijaya, Vishal, INDAM-1967, INDAM-7215, INDAM-62, INDAM-312, Varsha Improved, Vijya, Vishal, Pusa Sawani, Pusa Makhmali, Pusa A4, Hissar Unnat, Selection 1, Perkins Long Green, Parbhani Karanti,

Punjab Padmini, Punjab 7, Punjab 33, Punjab 13, Harbhajan, HB 55, Varsha Uphar, Vaishali Badhu, IHR 20-31, Arka Abhay, Arka Anamika.

EGGPLANT/BRINJAL : Kalia, Badshah, Green Long, Black Long, Saurabh L-13, Sufal, Suchitra, (H-47) Supriya R–14, 1632, 19794, Black Beauty, PRC, RB-63, Mysore Green Long, Pusa Purple Long, Pusa Karanti, Pusa Purple Cluster, Pusa Bhairav, Pusa Anupam, Pusa Anmol, Hissar Shyamal, Arka Navneet, DBSR 44, DBR 31, DBR 8, PH 4, Kalyanpur Type 3, Pant Samrat, Pant Rituraj, Benaras Giant, NDB 2, NDB 25, Punjab Barsati, Utkal Kesari, Utkal Madhavi.

References

- **A. Loughton** – 1998. "Production and Handling of Broccoli" ; Agriculture and Rural – Canada.
- **Crisp, P., Tapsell, C.R.** – 1993 "Genetic Improvement of Vegetable Crops"; Pergamon press, Oxford, U.K.
- **Ohiulips, R., Rix, M.** – 1993 "The Random House Book of Vegetables" Random House NY- USA.

REFERENCES

1. Report on Herbal Industry by Dr. V. Prasanna Bhat, MD, ITCOT 50-A, Graems Road, Chennai – 600 006.
2. The Macdonald Encyclopaedia of Medicinal Plants by Roberto Chiej, (1988), 66-73 Shoe Lane, London EC4P 4AB.
3. Growing Vegetables and Herbs by Bryce L. Patterson, Brombacher Books, Richmond, California.
4. Herbs That Heal, Natural remedies for Good Health by H.K. Bakhru (1993)- Orient Paperbacks-Delhi.
5. Ayurvedic Cure for common Diseases by Dr. N. Anjneya Murthi, D.P. Pandey (1987)-Orient Paperbacks-Delhi.
6. The Marshal Cavendish, Encyclopaedia of Gardening (1983), 58 Old Crompton Street, London, W1V 5PA.
7. The Seed Catalogue, Thompson and Morgan (1995), Poplar Lane, Ipswich, Suffolk, England, IP8 3BU.
8. Burpee's Commercial Catalogue (1991), W. Atlee Burpee and Co. Warminster, PA 18974.

9. Herbs by James Underwood Crockett, Ogden Tanner, The Time- Life Encyclopaedia of Gardening (1977).

References

- Chayote- *Sechium edule* (Jacq.) Sw. by Rafael Lira Saade, International Plant.
- Genetic Resources Institute –1996.
- Engels, J.M.M. and C. Jaffrey. 1993. *Sechium edule* (Jacq.) Swartz. Pp. 246-248 in Plant Resources of South East Asia. No. 8.
- Dictionary of Economic Plants in India by Umrao Singh, A.M. Wadhwani, B.M. Johri. 1983. Indian Council of Agriculture Research New Delhi.
- Dictionary of Indian Agriculture by M.G. Kamath. 1992. Allied Publishers Limited, New Delhi.
- http://www.wholehealthmd.com.

References

1. Encyclopaedia of Gardening by Marshal Cavendish, London-New York. 1983.
2. Thompson and Morgan, Poplar Lane, Ipswich, Suffolk, England, IP8 3BU.
3. Home Gardening by P.P. Trivedi, ICAR, New Delhi-1983.
4. Globe Artichoke, Commercial Vegetable production Guides, Oregon State University-2002.
5. Thompson and Morgan, Poplar Lane, Ipswich, Suffolk, England, IP8 3BU (1995).
6. Encyclopaedia of Gardening, Marshal Cavendish London-Newyork-1983.
7. Commercial Vegetable Production Guides, Oregon State University. http://www.orst.edu/Dept/NWREC.
8. Growing Jerusalem Artichoke, NC State University, Jonathan R. Schultheis, Department of Horticulture Science –1999.
9. http://www.wholehealthmd.com.
10. Ornamental Horticulture in India by K.L. Chadha, B. Choudhary (ICAR Publication1986).

References

1. BOSWELL - Our vegetable Travelers, National Geographic, Aug. 1949.
2. THOMPSON and KELLY-Vegetable crops, 1957- p.187-88.
3. STURTVENT : Notes on edible plants, 1919. P. 72-73.
4. VAVILOV : Botanical-Geographical principals of selection (Translation-1946).
5. Websters New International Dictionary, 1957.
6. THOMPSON *et al.* - Asparagus culture. Farmers Bulletin. 1646, USDA-1963.
7. McGillivray –Vegetable production-1953. p.282.
8. Fresh Fruit and Vegetables unload Totals-1965, 1964, 1963-USDA.
9. G.R. EHLERT and R.A. SEELIG-Fruit and Vegetable facts and Pointers, United Fresh Fruit and Vegetables Assoc.-Oct. 1966.
10. Agri. Hand Book-291, USDA-Losses in Agriculture-1965. p 87-90.
11. KAREN SCHLOEMAN-RAP Post Harvest Information Bulletin No.5-1995, Bethesda, MD, 20814 USA.
12. Germany. Export Manual for tropical fruits and vegetables: Asparagus. GTZ - Protrade. 27pp 1991.
13. King, G.A., Hurst, P.L., Irving, D.E. and Lill R.E. Recent advances in Post Harvest Physiology, Storage handling of Green Asparagus. Post Harvest News and Information. 4(3): 85-89. 1993.
14. Lipton, W.J. Post Harvest Biology of Fresh Asparagus. Hort. Reviews. 12:69-155.1990.
15. Falavigna, Agostina and Schiavi, Massimo. 7th International Asparagus symposium: Ferrara, Italy, June 19-23, 1989. Acta Hortic. Wageningen: International Society of Hort. Science. 1990 (271) 509 p.
16. Nichols, M.A. Asparagus production in tropics. Acta Hortic. Wageningen: International Society for Hort. Sciences. June 1992. (292)p. 149-153.
17. Pandita, P.N. and Bhan, M.K. (1992) Indian Horticulture., 36 (Oct-Dec): 23-26.
18. http://www.wholehealthmd.com.

References

1. Chutkaew, Chamnan. 1986. The future of Baby Corn. Proceedings of the 6th National Vegetable workshop. Nakhon Sri Thammarat, Thailand, pp. 11-20.
2. Chutkaew, C. 1986. Progress of Baby Corn Industry in Thailand. Paper presented in the FAO Expert consultation on Improved production of Tropical Vegetables. FAO regional office for Asia and the Pacific, Bangkok, Thailand.
3. Chutkaew, Chamnan; Paroda, R.S. 1994. Baby Corn Production in Thailand, A success story.
4. Asia Pacific Association of Agricultural Research Institutions, FAO Regional Office for Asia Pacific, Bangkok.
5. http://www.agro-on.com/a2market.htm (Baby Corn of the World).
6. Articles.

References

- http://www.wholehealthmd.com.
- http://www.organicgardening.com.
- NSW Agriculture Flemington Markets Reporting Service (Agdex 25/20).
- http://www.agric.nsw.gov.au.
- Luke Jewell Seeds Pty. Ltd. Australia.
- Sunland Seeds Pty. Ltd. Australia.

References

1. http://ohioline.osu.edu/hyg-fact/5000/5333.html.
2. http://www.ext.colostate.edu/pubs/foodnut/09330.html.
3. Sharon L. Mader, Extension Agent, Family and Consumer Sciences/4-H, Sandusky County.
4. How to Freeze the Right Way: Gerald D. Kuhn and Anne V.A. Resrricion. Cooperative Extension Service, The Pennsylvania State University.
5. So Easy to Preserve (Third Ed.) Janet Green, Ruth Hertzberg, Betrice Vaughn, The Stephen Greene Press, Lexington, Massachusetts.

6. Swami Vinayananda, 1984 "Dahlias of Today "Puget Sound Association of Washington.
7. Swami Vinayananda, 1985, "Dahlia Growing". Associated Publishing Company New Delhi.
8. Dr. Hardeep Singh, 2001 "Adhoonik Sabzi Vigyan", Published by Gramlok, 5/32, Patel Gali, Vishwasnagar, Shahdra, Delhi – 110 032.

References

- R.W. Robinson and D.S. Deckers – Walters (1997), "Crop Production Science in Horticulture".
- http://www.waynesword.palomar.edu/worthypl.htm.
- http://www.cucurbit.org.
- Peter Beckman – Golden State Bulb Growers, Moss Landing, California. Floraculture International - Dec. 2002.
- P.J. (Paul) van Leeuwen, International Bulb Centre "Seminar" – Floraculture International – Dec. 2002.

References

1. Harris, David R. 1972. **The origins of agriculture in the tropics.** American Scientist 60(2):180-193.
2. Hawkes, J.G. 1970. **The origin of agriculture.** Econ. Bot. 24:131-133.
3. Lowenberg, M.E., E.N. Todhunter, E.O. Wilson, J.R. Savage and J.L. Lubawski 1974. 2nd ed. **Food and Man**. Chapter 1. J. Wiley and Sons, NY.
4. Mangelsdorf, P.C., R.S. McNeish, and W.C. Galinat. 1964. **Domestication of corn**. Science 143:538-545.
5. Sauer, C. O. 1969. **Agricultural origins and dispersals**. Mass Inst. Tech. pp. 19-83, let-46.
6. Schwanitz, Fr. 1966. **The origin of cultivated plants**. Harvard Press, Cambridge, Mass (Good reference for kinds of changes accompanying domestication of plants).
7. Smith, C.E., Jr. 1969. **From Vavilov to the Present–A Review** - Econ. Botany 23.
8. Wilsie, C.P. 1962. **Crop adaptation and distribution**. Freeman, San Francisco. Chap. 5.

9. Bailey, L.H. 1949. **Manual of Cultivated Plants**. The McMillan Co., New York.
10. Richard, L.A. (editor) 1954. **Diagnosis and Improvement of Saline and Alkali Soils**. U.S.D.A. Handbook No. 60.
11. Rubatzky, V.E. and M. Yamaguchi. 1997. Chapters 2, 3, and 4 **World Vegetables.** 2nd Ed. Chapman and Hall, New York.
12. Smith, P.G. and J.E. Welch. 1964. **Nomenclature of vegetables and condiment herbs grown in the United States**. Amer. Hort. Sci. Proc. 8 Gre 4 : 535-548.
13. Gilmour, Chairman. **International Code of Nomenclature of Cultivated Plants - 1969**. International Bureau of Plant Taxonomy and Nomenclature Utrecht, Netherlands.